Public Policies for Environmental Protection

Public Policies for Environmental Protection

Paul R. Portney, *editor*
Roger C. Dower
A. Myrick Freeman III
Clifford S. Russell
Michael Shapiro

Resources for the Future • Washington, D.C.

Printed in the United States of America

Published by Resources for the Future
1616 P Street, N.W., Washington, D.C. 20036
Books from Resources for the Future are distributed worldwide by The Johns Hopkins University Press.

Library of Congress Cataloging-in-Publication Data

Public policies for environmental protection

Includes bibliographical references.
1. Environmental policy—United States. 2. Environmental law—United States. 3. Pollution—United States. 4. Pollution—Law and legislation—United States.
I. Portney, Paul R. II. Dower, Roger C.
HC110.E5P83 1990 363.7'056'0973 89-24363
ISBN 0-915707-52-7
ISBN 0-915707-53-5 (pbk.)

This book is the product of the Center for Risk Management at Resources for the Future. It was edited by Samuel Allen, designed by Debra Naylor, and indexed by Florence Robinson.

∞ The paper in this book meets the guidelines for permanence and durability of the Committee on Production Guidelines for Book Longevity of the Council on Library Resources.

RESOURCES FOR THE FUTURE

RESOURCES FOR THE FUTURE (RFF) is an independent nonprofit organization that advances research and public education in the development, conservation, and use of natural resources and in the quality of the environment. Established in 1952 with the cooperation of the Ford Foundation, it is supported by an endowment and by grants from foundations, government agencies, and corporations. Grants are accepted on the condition that RFF is solely responsible for the conduct of its research and the dissemination of its work to the public. The organization does not perform proprietary research.

RFF research is primarily social scientific, especially economic. It is concerned with the relationship of people to the natural environmental resources of land, water, and air; with the products and services derived from these basic resources; and with the effects of production and consumption on environmental quality and on human health and well-being. Grouped into four units—the Energy and Natural Resources Division, the Quality of the Environment Division, the National Center for Food and Agricultural Policy, and the Center for Risk Management—staff members pursue a wide variety of interests, including forest economics, natural gas policy, multiple use of public lands, mineral economics, air and water pollution, energy and national security, hazardous wastes, the economics of outer space, climate resources, and quantitative risk assessment. Resident staff members conduct most of the organization's work; a few others carry out research elsewhere under grants from RFF.

Resources for the Future takes responsibility for the selection of subjects for study and for the appointment of fellows, as well as for their freedom of inquiry. The views of RFF staff members and the interpretation and conclusions of RFF publications should not be attributed to Resources for the Future, its directors, or its officers. As an organization, RFF does not take positions on laws, policies, or events, nor does it lobby.

Contents

Tables

Figures

Foreword

Polls taken in the United States in recent years consistently indicate that the state of the environment is among the top public concerns. The public has also shown substantial support for using regulation to ensure environmental improvements, as well as an interest in the use of economic measures to ameliorate or correct perceived environmental problems. As administrator of the Environmental Protection Agency from 1985 to 1989, I encountered first-hand—and in all their complexity—those environmental problems that have aroused public concern, as well as some problems of which the public is just beginning to be aware. And in the search for solutions, I found that economic data essential to making wise policy decisions and regulations were all too frequently lacking.

There is no doubt that economics can play a vital role in improving environmental regulation. The use of economics means paying attention to costs as well as benefits, and it means making economic choices in deciding whether, when, and how to regulate. This book by Paul R. Portney and his coauthors—among the nation's foremost environmental policy analysts—addresses the economic issues involved in major areas of environmental regulation and policy. With directness and clarity, the authors examine the decisions that have been taken and their effects on regulation, policy, and environmental improvement. But the book looks forward as well as to the past, exploring the decisions that must be made today, the trends that are likely to affect environmental regulation tomorrow, and the need to face the tradeoffs between environmental protection and economic considerations.

In a word, this is a book for the public as well as the specialist—easily accessible to the general reader, and offering expert analysis and recommendations for the specialist. It provides important insights and understanding for those in government and out who take seriously the public's concern for environmental protection.

Lee M. Thomas
Chairman and Chief Executive Officer
Law Environmental, Inc.

Kennesaw, Georgia
October 1989

Acknowledgments

Any book requiring as much time to produce as this one inevitably depends on the contributions of many whose names do not appear on the cover. That is certainly the case here. Along with Resources for the Future, the Andrew W. Mellon Foundation provided the financial support that made this book possible. Tina Martin, Betty Cawthorne, and John Mankin patiently typed draft after draft of each of the chapters, several of which Christine Mendes carefully proofread. Marilyn Voigt quite indispensably took over the coordination of revisions, fact-checking, and other chores at a particularly trying time during the gestation period, for which I am most grateful. Sam Allen provided most skillful editorial assistance; Joan Engelhardt helped speed the design and production process; and Celia McEnaney kept the vision of the final product—a book people would read and learn from—firmly in my mind. John F. Ahearne played a critical role in the development of this book. From the very start, he alternately pushed, encouraged, commiserated with, and assisted me—always in the right way and at the right time.

During the period over which this book was conceived, written, and edited, more and more of my time came to be devoted to the inevitable phone calls, meetings, and trips. Simply put, this meant that the book increasingly came to be a nights-and-weekends project. Thus my most profound debt of thanks must go to my wife Susan. She not only tolerated this encroachment but also read drafts of several of the chapters to ensure that they were indeed accessible to non-economists. For her unflagging support on this and many other undertakings, I happily dedicate this book to her.

Washington, D.C.
November 1989

Paul R. Portney
Vice President, Resources for the Future,
Director, Center for Risk Management

About the Authors

Roger C. Dower is chief of the Energy and Environment Unit at the Congressional Budget Office. Before that he was research director at the Environmental Law Institute.

A. Myrick Freeman III is professor of economics at Bowdoin College, and a senior fellow in the Quality of the Environment Division at Resources for the Future.

Paul R. Portney is vice president of Resources for the Future and director of its Center for Risk Management. Formerly he was senior staff economist of the Council on Environmental Quality, Executive Office of the President, and visiting professor at the Graduate School of Public Policy, University of California, Berkeley.

Clifford S. Russell is professor of economics and director of the Vanderbilt Institute for Public Policy Studies. He was formerly a senior fellow and director of the Quality of the Environment Division at Resources for the Future.

Michael Shapiro is director of the Economics and Technology Division in the U.S. Environmental Protection Agency's Office of Toxic Substances. Before joining the EPA in 1980, he taught environmental policy in the John F. Kennedy School of Government at Harvard University.

one

Introduction

Paul R. Portney

The fall of 1977 saw the finishing touches put to *Current Issues in U.S. Environmental Policy*, the predecessor to this book. The introduction to that volume identified several factors that favored a thorough review of domestic environmental policy. These factors included the slow pace of environmental improvements following the legislation of the early 1970s, the high cost of complying with those laws, the arrival of a new president (Jimmy Carter) with the promise of a new approach to environmental regulation, the changing nature of environmental problems, and the then-current energy "crisis."

Although twelve years have elasped since then, these same reasons—and several new ones as well—justify still another in-depth review of the environmental policies of the United States. For instance, air and particularly water quality continue to improve quite slowly (if at all) in many places, even after a dozen more years of pollution-control spending by households, municipalities, and industrial polluters. This pace is not unrelated to the spotty record of compliance with air and water regulations and the sometimes feeble enforcement efforts that have been taken. Perhaps more distressing, very little progress seems to have been made in regulating hazardous wastes and toxic substances, even though the basic laws thought necessary to do so have been in place since 1976.

Nor has environmental regulation ceased being expensive. For example, according to the most recent estimate of the U.S. Environmental

Protection Agency (EPA), the cost of complying with federal air and water pollution-control regulations was $52 billion in 1981 alone (in 1988 dollars)—exclusive of the costs of pesticide, toxic substance, or hazardous waste regulations.[1] Between 1981 and 1990, according to the EPA, the nation will spend more than $640 billion in pursuit of clean air and water.

Since the earlier book, moreover, still another president has come to and gone from Washington, one with very different ideas than his predecessors about the means and ends of environmental regulation. Such change has not been without controversy. However, in one area the Reagan administration built upon and attempted to extend one of the major redirections in environmental policy charted by the Carter EPA, and it appears that President Bush will pursue this approach more vigorously than Reagan (although it is still too early in the Bush administration to be sure). Involved is the expanded use of economic incentives as supplements for or complements to more traditional types of regulation. This development merits explanation and discussion.

The focus of environmental policy also appears to be shifting once again, although in a more subtle way than before. To be sure, public health is still the primary emphasis of this nation's environmental laws. That much has not changed. Yet, an evolution is taking place in the kinds of health threats the EPA is addressing. The agency now spends much more time worrying about, and devotes more resources to, the toxic substances that pose longer-term and more exotic threats to health than it does to the so-called conventional air or water pollutants with which it was concerned for the first decade or so of its existence. While this apparent shift in emphasis is understandable, particularly given heightened public concern about hazardous substances and the risks they pose, it may also be regrettable if, collectively, we have an exaggerated view of the risks associated with the new breed of environmental hazards and at the same time turn our backs on older, now less "sexy" environmental problems. Some evidence presented below suggests that this *may* be the case.

Finally, there are new reasons to take a fresh look at U.S. environmental policy. At least one major piece of new legislation—the Comprehensive Environmental Response, Compensation, and Liability Act of 1980 (better known as the Superfund)—has been passed and significantly amended, and several of the older laws have been amended in important ways. In addition, a number of new environmental problems or policy issues that went unaddressed in the earlier volume have arisen or become more apparent. For all these reasons, it is again appropriate to review the current state of U.S. environmental policy and how it might evolve. That is precisely the purpose of this book.

The plan of the book is as follows. In chapter 2, I review briefly the evolution of federal regulatory policy from its early concern with natural monopoly, deceptive trade practices, and financial soundness toward its more recent preoccupation with environmental, safety, and health regulation. It was in the latter context that the EPA was created. After tracing briefly the growth of the EPA and the impact of its regulations, I turn to the sorts of choices that must be made in deciding whether and how to regulate, using examples illustrative of environmental policy in the United States. Chapter 2 should set the stage for the discussions that follow of the specific regulatory programs that protect the air, water, and land.

In chapter 3, I address the air pollution control policies of the United States, beginning with a review of efforts before 1970, when the Clean Air Act was amended to form the current basis for regulation. That review suggests that previous efforts were concentrated almost exclusively at the state and local levels, but that this began to change with the first federal laws in the 1950s. Following that discussion, I outline the major features of the Clean Air Act and several of their implications. Next I present some evidence on the effect of the act on air quality in the United States—this is, after all, the ultimate measure of effectiveness and one in which we should all be interested. The review suggests that, on the whole, air quality continues to improve in the United States.

Next I turn to an economic evaluation of the Clean Air Act. There I review selectively a number of studies providing evidence about the benefits and costs associated with air quality regulations under the act. Perhaps not surprisingly, one conclusion I draw is that it is no simple matter to compare the different estimates—they often pertain to slightly different views of the program, use different base years, or vary in other important ways. I conclude that both costs and benefits are likely to be substantial, but that saying more than this requires much care. Less ambiguous is another conclusion: it is possible to reduce the costs of meeting our current air quality goals quite substantially. A large number of studies support this conclusion. A final section presents some conclusions and policy recommendations.

In chapter 4, A. Myrick Freeman analyzes federal efforts to control water pollution in the United States. He too begins with a review of efforts prior to the substantial change that Congress made in 1972, a change that involved the goals, methods, and federal responsibilities in water pollution control. Next Freeman discusses in some detail the framework that grew out of the 1972 amendments to the Federal Water Pollution Control Act (now known simply as the Clean Water Act).

Basically, the approach in this framework differs from the environmental-quality-based approach in the Clean Air Act: the Clean Water Act, Freeman finds, concerns itself more with technology-based standard-setting and the use of federal grants to encourage waste water treatment by municipalities.

Freeman then turns his attention to the accomplishments under the new federal approach. He finds that these are difficult to identify because there is no single and unambiguous indicator of water quality, particularly at the national level. The problem is made more difficult by our inability to distinguish between changes in water quality that are a result of the act and those that might have occurred for other reasons. Freeman then reviews fragmentary evidence bearing on EPA's performance in carrying out the Clean Water Act, the performance of regulatees in meeting their responsibilities, and data on water quality since passage of the act.

Turning to the economic implications of the Clean Water Act, Freeman presents evidence on the costs and benefits associated with the act. In spite of uncertainty about the estimates, Freeman finds that the former are likely to exceed the latter, apparently by a considerable margin. He then suggests several kinds of changes that might be made in the act or in the way it is implemented that would improve this unfavorable picture. He devotes special attention in this discussion to the construction grants program and the control of non-point source pollution.

Roger C. Dower addresses current U.S. policy regarding the regulation of hazardous wastes in chapter 5. He begins by attempting to identify the scope of the problem—no simple matter given the lack of a standard definition of hazardous wastes and the absence of solid data on the number of abandoned disposal sites or the frequency with which such wastes turn up in drinking water or other environmental media. Following this introduction, Dower describes the two major laws passed to address the problems caused by hazardous wastes. In subsequent sections Dower discusses the economic impacts—both favorable and unfavorable—associated with the laws, the question of liability for damages caused by disposal sites (and the difficulty of securing insurance against such liability), the financing of cleanups at disposal sites, and the question of where the by-products of industrial production and household consumption will go if land disposal is ultimately prohibited.

In chapter 6, Michael Shapiro takes up the control of toxic substances by the Environmental Protection Agency. He begins by detailing the growth of the chemical and pesticide industries in the United States, particularly since the end of the Second World War. This discussion is useful because of widespread concern that the problems associated with toxic substances are due directly to these industries. Shapiro then explains the structure of the two major legislative responses to this perceived problem, the Toxic Substances Control Act (1976) and the Federal Insecticide, Fungicide, and Rodenticide Act (1947, with subsequent amendments).

Shapiro then identifies several major issues that arise in regulating pesticides and toxic substances. These include the information base upon which action must build, the speed with which the program must proceed, the definition of "unreasonable" risk and the way risks of this kind are to be balanced against benefits, and the problem of inhibiting the introduction of new and useful chemical products. In the course of this discussion, Shapiro presents what fragmentary evidence is available on the costs and benefits of toxic substance regulation.

In chapter 7, Clifford S. Russell reviews evidence on an often-ignored aspect of environmental policy in the United States—the ways our major laws are monitored and enforced. He begins by discussing the various forms of monitoring and enforcement, distinguishing between initial and continuing compliance, and between the monitoring of ambient quality and the monitoring of emissions by sources or other source activities. He then presents what sketchy evidence there is on the frequency of monitoring and inspection of polluters, and the kinds of enforcement actions that are taken, as well as their severity. One finding is that much more attention has been devoted to ensuring that polluters get pollution equipment in place than to seeing that it is operating correctly. Russell then examines several recent efforts to improve the monitoring and enforcement of environmental regulations. He concludes by suggesting additional measures that might be taken to strengthen this neglected but nevertheless quite important aspect of U.S. environmental policy.

In chapter 8, I draw some overall conclusions about progress in environmental protection in the United States, and identify several recent trends that are likely to shape the future direction of environmental policy.

NOTES

1. Environmental Protection Agency, "The Costs of Clean Air and Water Report" (1984).

two

EPA and the Evolution of Federal Regulation

Paul R. Portney

By comparison with many other federal regulatory agencies, the EPA and its siblings of the 1970s are relative newcomers. Indeed, it was more than a century ago, in 1887, that Congress created the Interstate Commerce Commission to regulate surface transportation industries, and it has been more than seventy years since the Federal Reserve Board and Federal Trade Commission were created to regulate commercial banks and deceptive trade practices, respectively. The first great burst of federal regulatory activity took place in the 1930s, during which Congress created the Federal Power Commission, the Food and Drug Administration, the Federal Home Loan Bank Board, the Federal Deposit Insurance Corporation, the Securities and Exchange Commission, the Federal Communications Commission, the Federal Maritime Commission, and the Civil Aeronautics Board.

Between 1938 and 1970 little took place in the way of new federal regulatory activity.[1] Following this lull, however, came the second major burst of federal regulation. In quick order were created the EPA, the National Highway Traffic Safety Administration, the Consumer Product Safety Commission, the Occupational Safety and Health Administration, the Mining Safety and Health Administration, the Nuclear Regulatory Commission, the Commodity Futures Trading Commission, and the Office of Surface Mining Reclamation and Enforcement. Along with the Food and Drug Administration (FDA) (and omitting the Commodity

Futures Trading Commission), these have come to be known as the "social" regulatory agencies. Generally speaking, they are those having to do with environmental protection and the safety and health of consumers and workers.

Aside from age, there are at least three important distinctions between the "old-line" agencies and the newer, social regulators like the EPA.[2] The first has to do with their reasons for being. In principle at least, all regulatory agencies have been created to remedy a perceived failure of the free market to allocate resources efficiently. Except for the FDA, the older agencies were meant either to control "natural monopolies" or to protect individuals from fraudulent advertising or unsound financial practices on the part of financial intermediaries or depository institutions.[3] (The latter justification was clearly a reaction to the calamitous Great Depression, during which many of these agencies were created.)

Federal intervention in the areas of environmental protection and the safety and health of workers and consumers has a very different rationale. Here, it is argued, the government must intervene because of externalities or imperfect information. The former arise when the production of a good or service results in some costs (like pollution damage) which, in the absence of regulation, are unlikely to be borne by the producer. In such cases, the prices of products will not reflect what society must give up to have them, so that Adam Smith's "invisible hand" will steer us awry. In the case of imperfect information, workers or consumers may be only dimly aware of the health hazards associated with various occupations or consumer products or foodstuffs. If so, they will be unable to trade off higher risks for either higher wages or lower prices in an informed way, so that the unaided market would not necessarily result in either the right amount or the correct distribution of risk.

The newer regulatory agencies differ from their elders in another important way, having to do with the specificity of their focus. The older agencies can be thought of as dealing with a single industry—the Interstate Commerce Commission (ICC) with surface transportation, the Federal Communications Commission (FCC) with communications, the Civil Aeronautics Board (CAB) with airlines, and so on. This is not true of the newer social regulatory agencies. Thus, for instance, the Occupational Safety and Health Administration (OSHA) regulates workplace conditions in a wide variety of industries ranging from chemicals to agriculture. Similarly, the EPA regulates emissions of air and water pollution from the electric utility, steel, food-processing, petroleum-refining, and many other industries. The broader mandate of the social regulatory agencies may be a more difficult one to satisfy, since it requires each agency to become knowledgeable about and sensitive to the special problems and production technologies in many different industries.

A final distinction between the old-line agencies and their newer counterparts concerns recent developments in the scope of their activities. While it is always difficult to generalize over members of such a diverse group, the thrust of recent legislation and administrative actions at the older regulatory agencies has been in the direction of a sharp curtailment in the extent of their intervention in the markets they have regulated. This has been most pronounced in the case of the CAB, which was legislated out of existence in 1985 after it was recognized that the airline industry was ripe for competition. Similarly, the FCC has proposed significant reductions in the scope of its own activity, and both the trucking and railroad industries have been substantially deregulated through recent legislation and through administrative rulemakings at the ICC. The financial regulatory agencies and the Federal Energy Regulatory Commission (FERC, the successor to the Federal Power Commission) have also seen their powers eroded over time.

In general, this has not been the case at the newer, social regulatory agencies. While many have questioned the *way* these agencies have pursued their goals, relatively few have suggested that there is no need for them at all. It would be difficult to argue that unfettered competition among firms would lead to the right amount of pollution, product safety, or workplace risk so long as the problems of external effects or imperfect information characterize the conditions of production or consumption. Thus, while there is much agitation for regulatory reform—about which more is said below—there are very few calls for the abolition of the EPA, OSHA, FDA, or other social regulatory agencies. Indeed, in the case of the EPA at least, new responsibilities and regulatory programs have been added to the old almost continually since its creation. Trying to initiate new programs while struggling to master the existing ones is one of the problems with which the EPA has had to contend constantly.

THE CREATION AND GROWTH OF THE EPA

At the vanguard of the new social regulatory agencies, the Environmental Protection Agency got its start on July 9, 1970 when President Nixon submitted Reorganization Plan No. 3 of 1970 for congressional approval. That reorganization plan proposed to consolidate under one roof—the EPA's—various functions being performed at that time by the departments of Interior, Health, Education and Welfare, and Agriculture, as well as by the Atomic Energy Commission, the Federal Radiation Council, and the Council on Environmental Quality. By December 1970 the plan had been approved by Congress and the EPA was in action.

Because it was created out of existing programs, the EPA was never a very small agency. In 1971, its first full year of existence, it had about 7,000 employees and a budget of $3.3 billion, $512 million of which went to operate the agency, with the remainder being passed through the EPA in grants to state and local governments. By 1980, the agency had grown to more than 12,000 employees; its budget by that time was more than $5 billion, $1.5 billion of which went to the operation of the agency itself.

The budget of the Environmental Protection Agency, like that of most federal agencies, shrank during the Reagan years. In its final budget submission, just before George Bush was sworn in as president, the Reagan administration requested $4.8 billion for the EPA for fiscal year 1989—$1.6 billion for operations, $1.5 billion for sewage treatment grants, and $1.6 billion for the Superfund (in 1989 dollars). Ignoring the latter program, which did not exist in 1980, and adjusting for inflation, the operating budget of the EPA has fallen by about 15 percent in real terms since 1980.

One must be careful not to ascribe too much importance to the budget of the Environmental Protection Agency (or any other regulatory agency). While the budget may provide some guidance as to the agency's capabilities, its spending authority is less important than the costs incurred by those subject to the agency's various regulations. The latter, called compliance costs by economists, never show up in the agency's budget, yet they can dwarf its operating costs. For example, in 1981 the EPA spent $1.8 billion on outside research, salaries, rent, and other operating expenses (expressed in 1988 dollars). Yet in that same year, as indicated above, those subject to the EPA's air and water pollution control regulations were forced to spend some $52 billion to comply with EPA requirements. These expenditures—for pollution control equipment, cleaner fuels, sludge removal, additional manpower—give a much more accurate picture of the economic importance of the EPA than does its budget. (As later chapters point out, environmental regulations may of course result in substantial benefits as well. One must look at both costs and benefits before passing judgment on the overall worth of a particular regulatory program.)

The rapid growth of off-budget environmental compliance costs (they were estimated to be only $29 billion in 1973) has led to periodic concern about the possible effects of pollution control spending on the overall performance of the economy. While this is not the place to review in any detail the many studies on this subject, it may be useful to summarize them briefly.[4] With a fair degree of consistency, these studies have found that pollution control spending has had a relatively minor impact on macroeconomic performance. It has exacerbated inflation somewhat and slowed the rate of growth of productivity and of the GNP. On the other hand, studies have found that pollution control spending appears to have provided some very modest stimulus to employment, at least during the time when spending for sewage treatment plant construction was high. None of this should be too surprising. While annual expenditures of $52 billion or more are hardly trivial, they are small in comparison to a $5 trillion GNP like that of the United States. Thus it is unlikely that environmental or other regulatory programs on the present scale will ever be found to exert a significant impact on the measured performance of the economy.[5]

FUNDAMENTAL CHOICES IN ENVIRONMENTAL REGULATION

It may prove helpful in understanding the following chapters to review some of the basic choices that must be made in environmental regulation. The first question is one whose answer is often taken for granted—whether to regulate at all. While today we tend to take the need for regulation as a given, there are several possible alternatives to environmental regulation. These include the private use of our legal system as well as private negotiation or mediation.

Both alternatives depend upon a clear prior definition of property rights. Imagine that it was clearly understood that any citizen had an absolute right to be compensated fully for the damages from any kind of pollution. If the smoke from a neighbor's wood stove or a factory were ruining your laundry business, you could take the offending party to court. If your damages could be accurately assessed, and if the polluter were held liable for them, the legal approach would create the right incentives for polluters. They would undertake certain pollution control measures if the costs of these measures were less than the damages they would have to pay you. And they would continue to emit some pollution—and pay you for the damage it does—if it were more expensive to control than it was to reimburse you. Economists like such solutions because they minimize the total costs—control expenditures plus residual damages—associated with pollution control. Until the start of the twentieth century, all air pollution problems were handled under the nuisance and trespass provisions of common law.

Unfortunately, the real world is much more complicated than this simple example would suggest, and the added complexity makes a purely legal approach to environmental protection much less practical. First, it is not perfectly clear where property rights in clean air are, or even ought to be, vested in our society. This may seem puzzling, since most people accept the right of the citizenry to be free from pollution. Yet even this apparently sensible proposition seems strained in, say, the case of a laundry that deliberately moves from a clean location to one directly adjacent to a factory and then demands compensation for smoke damage. In other words, it seems to make some difference who was there first. Furthermore, if property values and rents are lower in polluted areas than in clean ones (as they generally are), the launderer seems to be on shakier ground still, since he would have already reaped some savings in operating costs (lower rent) by virtue of his new location. In one sense, he would be getting double benefits if the factory were forced to curtail its operations. Not surprisingly, arguments like these are often advanced in defense of firms resisting pollution control. They are not wholly without merit.[6]

Even if property rights were clearly defined, environmental protection via the legal system or private negotiation would not be without difficulties. For instance, pollution rarely occurs on a one-to-one basis as in the simple example above. There are often many polluters, thus making it difficult or impossible to know which factory, car, or wood stove is responsible for which damages. Also, there are generally many "pollutees," no one of whom may be suffering sufficient damages to merit taking legal action or initiating mediation or negotiation alone. Legal transaction costs may be so high that they inhibit the filing of class action suits, even though aggregate damages across all pollutees may be significant. Finally, some of the damaging effects of pollution may be both more subtle yet more serious than mere dirty laundry. For example, pollution may be one of the causes of cancer and other serious illnesses. Yet the long latency period between exposure and manifestation, coupled with the possibility of other causes, will make it virtually impossible to assess liability satisfactorily in a courtroom or arbitration chamber. Add to this the difficulty of valuing the pain and suffering from such illnesses, and one can quickly understand the possible shortcomings of alternative approaches to government intervention.

Having said this much, it is time to make one more quite important point. Proponents of government action, regulatory or otherwise, are quick to point as justification to the imperfections inherent in free markets. Public goods, externalities, natural monopolies, and imperfect information—these are all problems economists recognize as standing in the way of efficient resource allocation. Yet regulation seldom goes exactly as planned when it is substituted for the forces of the market.[7] It is often poorly conceived, time-consuming, arbitrary, and manipulated for political purposes completely unrelated to its original intent. Thus the real comparison one must make in contemplating a regulatory intervention is that between an admittedly imperfect market and what will inevitably be imperfect regulation. Until it is recognized that this is the dilemma before us, we will be dissatisfied with either approach.[8]

If government intervention *is* deemed desirable, one must then ask, at what level should intervention take place? While environmental protection is very important, so too are public school quality, police and fire protection, income assistance, and criminal justice. Yet the latter are all functions which, in our federal system, are entrusted largely to local or state governments. Thus it is not obvious that all or even most environmental regulation should take place at the federal level. This is a key decision that must be made in designing interventions, and as chapter 7 points out one recent trend is in the direction of much more state and local regulatory activity.

Once the decision has been made to intervene at some level of government, the next choice to be made is, how should we decide how much protection to provide? There are a number of frameworks for deciding the answer, and Congress has directed different social regulatory agencies to use different frameworks in establishing levels of protection.[9] In fact, even within the EPA the approach differs depending on the regulatory program in question.

One way to select the degree of protection might be called the zero-risk or safe-levels approach. The administrator of the Environmental Protection Agency would be directed to set a particular environmental standard at a level that would ensure against any adverse health (or other kind of) effect. As later chapters illustrate, this approach is not uncommon. And on its face, it certainly seems reasonable. After all, would we want a standard to be set at a level that poses some recognizable threat to health? Surprisingly, perhaps, the answer is maybe.

Science and economics contribute to this unexpected response. Accumulated research in physiology, toxicology, and other health sciences suggests that for a number of environmental pollutants, particularly carcinogens, there may be no threshold concentrations below which exposures are safe. This implies that standards for these pollutants must be set at zero concentrations if the populace really is to be protected against all risks. Here economics intrudes in a jarring way. Simply put, it is impossible to eliminate all traces of environmental pollution without at the same time shutting down all economic activity, an outcome which neither the Congress nor the public would abide. Yet this is where the zero-risk framework often appears to lead if interpreted literally.

Having raised this disquieting possibility, let us push it further to consider another interesting case. Suppose a particular pollutant was harmful at ambient (or outdoor) levels to one very large group of people—the one-third of U.S. adults who choose to smoke cigarettes—but only this group. If the costs of reducing ambient concentrations of the contaminant were very large, might society not decide to forgo this health protection? It might well, in view of the role that the sensitive population has played in predisposing itself to environmental illness. Here too, then, the zero-risk approach would cause problems.

Perhaps more realistically, decisions to live with some risk might be reached even if the group at risk had done nothing to create its sensitive status. At some point, the costs of additional protection might be judged by society to be too great if the added health benefits are relatively small. Painful though such decisions may be, they are the rule rather than the exception in environmental and other policy areas. The problem with the zero-risk approach is that it prevents such tradeoffs from being made.

Another approach to the how-much-protection question is a variant of the above. It is often referred to as the technology-based approach. Under this framework, the only pollution permitted is that remaining after sources have installed "best available" or other state-of-the-art control technology. The underlying idea is simply that all technologically feasible pollution control measures will be required, and only after that will residual risks be accepted. This approach is somewhat weaker than a strict zero-risk approach since it admits the possibility of some risks. But in its strictest form, it is uncompromising with respect to trading off cost savings for less strict pollution control. For this reason, it appeals to many.

In its application, however, the technology-based approach faces several drawbacks. First, there is no unambiguously "best" technology—emissions can always be reduced further for additional control expenditures. In the limit, of course, sources could be closed down entirely—an ultimate, perhaps draconian, form of best technology. Moreover, implicit in the technology-based approach is the assumption that the control that results must be worth the cost. This might well depend upon the particulars. For instance, very strict control may be deemed essential for polluters in densely populated areas but much less important for those in remote, unpopulated regions. Yet the uniform technological approach has the liability of precluding the tradeoffs necessary to decide such questions. In addition, the technology-based approach suffers in a dynamic setting because it locks sources into specific means of control. It is unlikely that a firm required to meet this year's best technology will be told to scrap that equipment if next year's is even better. Thus this approach may deprive us of the opportunity to reduce the costs and increase the efficiency of pollution control over time.

The final framework for standard-setting discussed here formalizes the notion of balancing and incorporates it into environmental law. The relevant statutory language might direct the administrator of the EPA to set standards to protect health and other values while at the same time taking account of the costs and other adverse consequences of the regulations. The advantage of this approach is that it makes possible, indeed mandatory, the kinds of tradeoffs we have suggested might be desirable. On the other hand, it also forces the administrator to make very difficult decisions. Moreover, if all favorable and unfavorable effects are supposed to be expressed in dollars, so that precise benefit-cost ratios are required, this approach would impose a burden that economic analysis is not prepared to bear. In spite of the recent progress in valuing environmental benefits and costs,[10] the science is still far short of being able to make such comparisons in a precise way. For this reason, the balancing approach is best left in a qualitative or judgmental form.

At this point, some might reasonably chafe at the balancing framework. Why compromise citizens' health or welfare so that corporate or other polluters might remain economically healthy? This very natural question deserves a straightforward answer, one which proceeds along the following lines. While it would be nice if there were, there are no disembodied corporate entities into whose deep pockets we can reach for pollution control spending without at the same time imposing losses on ourselves or our fellow citizens. This is because corporations are merely legal creations, the financial returns to which all accrue to individuals in one capacity or another. Thus if corporations spend more for pollution control, these costs may be passed on to others in the form of higher product prices.[11] Alternatively, if costs cannot be shifted to consumers, then stockholders, laborers, or the management of the corporations will suffer reduced earnings.

Thus far the tradeoff may still seem appealing since it is expressed in terms of dollars versus health. However, the reduced incomes of consumers, stockholders, or employees will eventually mean less spending on goods and services they value. At this point, then, the tradeoff becomes more stark. Pollution control spending can sometimes protect health, but at an eventual cost to society of forgone health, education, shelter, or other valued things.[12] The real trick in environmental policy—or any other area of government intervention—is to ensure that the value of the resulting output is greater than that which must be sacrificed. And this sacrifice will take place regardless of the framework for standard-setting that is being employed. In the economists' view of things, the balancing approach is desirable not only because it is a natural way to make decisions, but also because it brings out in the open the terms of trade, so to speak. If we dislike the compromises being made by our regulatory officials, we can demand their removal.

Once environmental standards (or ambient standards, as they are sometimes called) have been selected, the next step is deciding on the means of attainment. In other words, how do we control the sources of pollution so that the environmental goals are met? While there are other possibilities, the two most common approaches are via direct, centralized regulation, or through an incentive-based, decentralized system.[13] Under the first, individual polluters are assigned specific emissions reductions; under the latter, they are given more latitude.

Under the centralized approach, the regulating authority has considerable discretion in apportioning the emission reductions required to meet the ambient standard. Only when there is but one source of pollution is it clear where emissions must be reduced.[14] More typically, there are multiple sources; in such cases the authority has to decide how much each source must curtail its offending activities. There are several ways this decision can be made.

If aggregate emissions must be reduced by 25 percent to meet the environmental standard, for example, each source could be required to cut back its own emissions by 25 percent. This equiproportional rule has the very attractive feature of *appearing* fair. Why only the appearance of fairness? Because of the very great diversity of sources for many environmental contaminants, ranging from neighborhood dry-cleaners or car-repair shops to complex steel mills or large chemical plants. The differing characteristics and technological circumstances of these sources mean that one source may be able to reduce its emissions by 25 percent quite inexpensively, perhaps by switching to a less polluting fuel or altering slightly its manufacturing technique. Yet another source might find that it can meet its 25 percent reduction only through the installation of expensive control technology. Thus a requirement for equal-percentage reductions may mean very unequal financial burdens.

Under another approach, emission reductions might be apportioned on the basis of affordability—that is, the largest cutbacks might be required of those in the best financial shape. This, too, has some obvious appeal. Indeed, under our present individual income tax system, we ask those in higher income brackets to pay a higher percentage of their income in taxes, and this would seem to extend that principle to pollution control.

On closer inspection, however, assigning emission reductions on the basis of ability to pay also has serious drawbacks. First, it would penalize successful, well-managed firms and reward laggards that may well be largely responsible for their own poor financial state. In this sense, then, the approach gives exactly the wrong set of signals to firms and slows the replacement of failing enterprises with newer, more efficient ones. Second, there may be no relation whatsoever between a source's emissions and its financial condition. Thus a very profitable firm may have very low emissions (particularly if it has continually modernized) but under this approach would still be forced to spend heavily on further emission reductions; meanwhile, a smoke-belching firm in perilous financial condition would be let off lightly. For these reasons, an affordability criterion is less attractive than it may at first appear.

Finally, the regulatory authority could try to apportion emission reductions among sources in such a way that the required aggregate reduction was accomplished at the least total cost to society. In other words, the central regulator could look across all sources and ask where the first ton of emissions might be reduced most inexpensively, then require it to be removed there. The second ton of emissions reductions would then be assigned, again to the source that could accomplish it most cheaply. And so on until the aggregate emissions goal had been met.

This approach has the advantage of ensuring that society (through the affected sources) gives up as little as possible to get the emission reductions. But it raises the possibility of another sort of inequity. Suppose that one source, among a large number of polluters in a particular area, was always the lowest-cost abater? This is unlikely to be the case in reality, but it might hold true in certain circumstances. It would hardly seem fair to place the entire burden of emissions control on that source merely because it could reduce pollution more inexpensively than the other sources. Thus, although the cost-minimization approach has some obvious appeal, it is not ideal.

Decentralized approaches, on the other hand, do address certain of these problems, although they present difficulties of their own. Perhaps the best-known of the decentralized approaches is the effluent charge or pollution tax.[15] Under this scheme, the regulatory authority imposes a tax or fee on each unit of the environmental contaminant discharged. In its purest form, the charge would be set to reflect the damage done by each unit of emissions. Rather than tell each firm how much to reduce emissions, the authority would leave it to the firm to respond to the charge however the firm best sees fit. Some sources would reduce their emissions immediately—those will be the ones that can do so at unit costs less than the amount of the charge. By doing so, they save the difference between their per-unit cost of control and the per-unit charge. Other sources will find it economical to continue discharging—they will be the ones finding it cheaper to pay the tax than to incur the required control costs.

Such an approach has several advantages. First, it ensures that the sources that do elect to take control measures are those with the lowest control costs. In other words, it mimics the least-cost approach under command-and-control, but does so without requiring the central authority to specify emission reductions for each and every source. Second, and perhaps more important, it provides a continuing incentive for firms to reduce their costs of pollution control. Since they must continue to pay the per-unit charge, it continues to be economical for them to find ways to reduce emissions for less than that charge. Third, this system requires something from all sources—either they must reduce pollution to escape the charge, or they must continue to pay the charge. No one gets off scot-free.

As might be expected, the effluent-charge route has shortcomings, which at least some economists have been slow to recognize or acknowledge.[16] For one thing, it is no picnic to determine the damage done by each unit of pollution; in practice this could only be approximated at best. Some have suggested that the difficulty of apportioning damage is such a liability of the charge approach that a modified version of the approach should be used.[17] Under this variant, the central authority would first select the desired level of environmental quality and would then set the charge at a level sufficient to induce the emissions control that would achieve it. Yet even such a variant would require some trial and error, and the uncertainty this might create could make firms reluctant to come to their initial emissions control decisions. The effluent-charge route also presents one serious political problem. Under this approach, the emissions that sources are free to discharge under the current permit system would be subject to the charge. Thus many sources that presently complain about over-regulation would have a new complaint: a major effluent-tax liability.

A second variant of the incentive-based approach involves marketable pollution "rights" or permits. This approach could work in one of two basic ways. Under one version, the central authority would first decide how much total pollution was consistent with the predetermined environmental goal. It would then print up individual discharge permits, the total quantity of which added up to the maximum amount permitted. No one without a permit would be allowed to discharge the regulated pollutants. The permits could be allocated among sources in one of several ways. First, a sale might be held at which all of the permits were auctioned off to the highest bidders. Alternatively, the permits could be distributed free of charge on some predetermined (or even random) basis—perhaps on the basis of historical levels of pollution. Either way, the permits would be marketable anytime after the initial distribution.

The incentive effect under a system of marketable permits is not unlike that of the effluent charge discussed above. Those sources that currently pollute but which could reduce pollution for less than the cost of a permit would take control measures. Those sources finding it very expensive to reduce pollution would buy discharge permits instead. Thus, as if guided by the same invisible hand, the emission reductions would take place at the low-cost sources, thereby minimizing the costs associated with a given reduction in emissions. Similarly, those firms buying permits would have a continuing incentive to reduce their costs of pollution control—as soon as they could do so, they could stop buying permits and save themselves money in the process.

The permit approach has one major advantage when compared to the effluent charge: the permit approach looks more like the existing system, which involves permits issued by the EPA or state environmental authorities, than does the latter. This may sound strange, but radical change is almost always more difficult to accommodate than gradual change. Since the marketable permit system is capable of accomplishing most of the same things as the effluent charge, why not advance it if it will be easier to put in place? This logic appears to have prevailed, and the inroads made by incentive-based approaches in environmental policy over the last ten years have featured permits.

Marketable permits are not without shortcomings, of course. One concern has to do with the possibility that certain sources might buy up all the permits as an anti-competitive tactic. While this ought to be rectifiable through governmental antitrust actions, in practice such actions might take time. Another question concerns the initial distribution of permits before the development of secondary markets. If all the permits are auctioned off, this approach would fall prey to the same political problems that arise under a charge approach—some sources would have to pay for emissions they are granted free under the existing system. Thus political problems could become formidable. If the initial permits are to be distributed free of charge, how should they be allocated? On the basis of previous emissions? To all citizens equally? To environmental and industry groups? This too is a potentially thorny problem, although not an insurmountable one.

A fifth and final question that arises in environmental policy is often overlooked: How do we monitor for compliance with the standards we set, and take enforcement actions against those in violation? We say less about this here because we devote an entire chapter to it. Suffice it to say now that the choice of both environmental and individual sources discharge standards ought to be (but often is not) influenced by the realities of monitoring and enforcement. Key issues involve the extent of reliance on financial penalties for noncompliance, the comparative strengths and weaknesses of civil as opposed to criminal penalties, and the choice of a monitoring strategy in a world of limited resources. All these issues and more must be addressed in designing sensible environmental policies.

U.S. ENVIRONMENTAL POLICY: A HYBRID APPROACH

As one might suspect, the fundamental questions raised above have been answered in an eclectic and hybrid way as environmental policy has evolved in the United States. With respect to the decision on intervention, federal, state, and local governments have all decided to intervene. Environmental statutes exist at all three levels—in fact, special districts have been formed in many areas around environmental problems. Thus long ago the decision was made not to entrust environmental problems and disputes solely to markets, to the courts, or to mediation services.

This observation also suggests the level at which intervention has taken place—at every level. Even federal environmental laws reserve important functions for state and local governments. For instance, under the Clean Air and Clean Water acts, the federal government (as embodied in the EPA) sets important ambient environmental and source discharge standards, yet the monitoring and enforcement of these standards is left largely to the states and localities. In fact, some ambient and source discharge standards are themselves reserved for lower levels of government in certain important cases. In other words, even federal laws are "federalist" in nature.

As to the choice of goals (How safe should we be?), U.S. environmental laws embody a range of approaches. A number of the most important environmental laws, or parts thereof, reflect the zero-risk (or threshold) philosophy. For instance, the Clean Air Act directs that ambient standards for common air pollutants be set at levels that provide an "adequate margin of safety" against adverse health effects, while standards for the so-called hazardous air pollutants are to provide an "ample margin of safety." Under the Clean Water Act, ambient water quality standards—which are left to the states rather than the federal government to establish—are also to include a margin of safety for the protection of aquatic life.

Other environmental standards are based on the technological approach to goal-setting. This is true of the Clean Air and Clean Water acts, the Resource Conservation and Recovery Act, and the Safe Drinking Water Act, as later chapters will discuss in some detail. The notion of best-available technology—along with its cousins "best-conventional" and "reasonably available" technology and "lowest-achievable emissions," as well as others—plays a big role in U.S. environmental policy, even in those statutes which in other places embrace the zero-risk goal.

Even the balancing framework favored by economists is alive and well in environmental policy. For although balancing appears to be prohibited under certain sections of the Clean Air and Clean Water acts, it is *mandated* under the most important parts of the Toxic Substances Control Act and the basic pesticide law, the Federal Insecticide, Fungicide, and Rodenticide Act. One is tempted to throw up one's hands and say, You figure it out! One of the purposes of the following chapters is to help the reader in doing just that.

In one important respect, the environmental laws have been rather uniform. When it comes to the means of pursuing environmental goals, the centralized or command-and-control approach has been given precedence over incentive-based approaches. Congress has rather consistently written regulations directing the EPA to establish emissions standards (with the help of the states and localities) and to issue and enforce permits specifying those standards. However, it is certain that a variety of factors have influenced the emission reductions the Environmental Protection Agency has required. While the agency often claims to have pursued a least-cost strategy, the uniform rollback and ability-to-pay criteria have clearly dominated in apportioning emission cutbacks.

The effluent charge approach has never really gotten off the ground in U.S. environmental policy, although a tax on emissions of sulfur into the air was proposed by the Nixon administration in the EPA's first year and once again in 1988. Recently, however, the standards-and-permits approach to air pollution control has evolved in the direction of marketable permits, and there is talk of applying this approach more widely in air pollution control as well as in other regulatory programs. Needless to say, this is an important development, and each of the chapters below touches on the possibilities for an expanded role for economic incentives.

Before concluding this chapter, some mention should be made of several generic problems that have arisen in the hybrid U.S. environmental policy since 1970. These problems have nothing to do with the well-known difficulties that arose at the Environmental Protection Agency between 1981 and early 1983;[18] rather, they have to do with the fundamental approach that Congress has taken in environmental regulation. A brief review of them may provide valuable perspective for the following chapters.

These problems are of four sorts.[19] The first has to do with the tremendous complexity of our environmental laws and their penchant for promising a very great deal in a very short time. For instance, the clean air and clean water laws promise "safe" air and water quality, call for the establishment of literally tens of thousands of discharge standards, mandate the creation of comprehensive monitoring networks, and impose numerous other important tasks on the administrator of the EPA. Yet the laws allotted just 180 days for completion of many of these responsibilities. Today, more than seventeen years after passage of the laws, many of those assignments have yet to be carried out.

Similarly, the Toxic Substances Control Act calls for the promulgation of separate testing rules for each new chemical. Yet although such chemicals come on the market at the rate of 1,000 per year, the EPA has issued testing rules for only a few substances. Of the more than 50,000 existing chemicals in commerce, only a small fraction have been tested for carcinogenicity or other harmful effects. Each of the environmental laws provides examples like this where Congress either misunderstood the time required to issue careful regulations, or disregarded it in the rush to get legislation on the books. The EPA has tried to run faster and faster since its creation, but has fallen farther and farther behind because of its impossible burden and an occasional lack of will.

Problems of the second sort have to do with the spotty compliance with those standards that have been issued, and our poor ability to know which standards are being violated and which sources are responsible. The reason for these problems are two, it would appear. First, monitoring both ambient environmental quality and the emissions from individual sources is much more complicated and expensive than one would imagine. Monitoring is not a straightforward matter, as might be supposed from reading the laws. Second, monitoring and enforcement have always been poor stepsisters in the eyes of Congress. Apparently it is more fashionable to write new laws and call attention to problems with existing laws than it is to engage in the dirty work of fashioning an enforceable and scientifically meaningful set of standards. Thus enforcement programs have always suffered financially at the expense of new and emerging regulatory programs.

The third sort of generic problem concerns the frequent emphasis in environmental statutes on absolutist goals. Waters are to be "fishable and swimmable" as one step toward a world of "zero discharges" into rivers, lakes, and the oceans. Conventional and hazardous air pollutants are to be at "safe" levels, as are drinking water contaminants. Such an approach has obvious political appeal—it is comforting to tell voters that they will be safe from all environmental threats. But that will simply not be the case unless standards are to be set at zero, an impossibility for most pollutants. Thus, although it is surely done, the balancing of environmental versus other important goals, economic and otherwise, is done implicitly. This has resulted in setting some standards at levels that appear to be hard to justify on any rational basis.

Finally, and perhaps inevitably, environmental statutes have become contaminated by redistributive goals which often work against the environmental programs in which they are nested—the fourth sort of problem. For example, newly built electric power plants are forbidden to reduce sulfur dioxide emissions by switching from high-sulfur to low-sulfur coal in order to protect the jobs of a small number of high-sulfur coal miners. This prohibition exists in spite of the tremendous cost savings that might be reaped if fuel-switching were permitted. Similarly, federal subsidies for the construction of sewage treatment plants have been continued even though the plants seem to have had a questionable impact on water quality in many areas, and although the federal subsidy has crowded out state and local spending for these same plants. Apparently, the pork-barrel aspects of the program have proved too attractive to eliminate. Water pollution from farms and other non-point sources has been overlooked altogether because of the political power of the parties that would be affected by tighter controls. Some evidence also suggests that environmental regulations have been structured to protect declining regions of the country from the effects of further economic growth in faster-growing sunbelt states. While all these contortions of environmental policy are understandable, they also stand in the way of an effective and less costly approach to environmental protection. As such, they deserve to be starkly highlighted.

NOTES

1. Exceptions were the creation of the Federal Aviation Administration, the Federal Highway Administration, and the Federal Railroad Administration to oversee safety in the respective industries.

2. See George C. Eads and Michael Fix, *Relief or Reform? Reagan's Regulatory Dilemma* (Washington, D.C., Urban Institute Press, 1984) pp. 12–15.

3. A natural monopoly is said to exist when the per-unit cost of producing a good or service continues to fall with increases in output. In such a case, it is argued, consumers will enjoy the lowest prices if one firm serves the whole market rather than sharing it with two or more competitors, as long as the single provider is regulated so as not to abuse its monopoly position. Natural monopolies are thought to arise most often when the fixed costs of doing business are a large proportion of total costs. The traditional examples are local telephone service, electricity distribution, and natural gas pipelines.

4. See Paul R. Portney, "The Macroeconomic Impacts of Federal Environmental Regulation," *Natural Resources Journal* vol. 21 (July 1981) pp. 459–488.

5. If regulation is the barrier to entry and discouragement to new growth that some maintain it is, and if these effects could be quantified, this conclusion could change.

6. In the real world, in fact, property rights appear to be shared. Even under existing regulation, firms are generally permitted to emit at least some pollution without being held responsible for the damage it may cause. In addition, tax breaks are provided for some investments in pollution control equipment. In effect, this shifts some of the costs of pollution control to the taxpayer, as would be the case if the property right were initially vested in the polluter.

7. See Charles Wolf, Jr., *Markets or Government* (Cambridge, Mass., MIT Press, 1988).

8. While it is probably inappropriate in the case of environmental problems, there is a third alternative to regulation in cases involving occupational hazards or dangerous consumer products. There the government could limit its role to the provision of information about the risks inherent in different jobs and/or products. Workers and consumers could then hold out for higher wages for risky jobs (as they do now) or pay low prices for risky products. Employers or producers would then have to decide whether it was in their interest to continue to pay higher wages or receive lower prices rather than reduce the hazards. In this way, too, market forces would work toward optimal riskiness. It should be noted, of course, that this model would be successful only if workers and consumers were fully informed about such risks and only if they had a range of jobs and products from which to choose.

9. See Lester B. Lave, *The Strategy of Social Regulation* (Washington, D.C., Brookings Institution, 1981).

10. See A. Myrick Freeman III, *The Benefits of Environmental Improvements: Theory and Practice* (Baltimore, The Johns Hopkins University Press for Resources for the Future, 1979); and Allen V. Kneese, *Measuring the Benefits of Clean Air and Water* (Washington, D.C., Resources for the Future, 1984).

11. This is the point of such regulations, in fact, since the higher prices discourage consumers from purchasing products whose production generates pollution.

12. Recently an effort was made to link spending like that necessitated by pollution controls to possible premature mortality via reductions in individuals' wealth. See Ralph L. Keeney, "Mortality Risks Induced by Economic Expenditures," working paper, University of Southern California, Systems Science Department (July 1988).

13. Other possibilities are moral suasion and direct government purchase of pollution control equipment. See William J. Baumol and Wallace E. Oates, *Economics, Environmental Policy, and the Quality of Life* (Englewood Cliffs, N.J., Prentice-Hall, 1979) pp. 217–224.

14. Even in this apparently simple case the matter is not so straightforward, because nature accounts for a share of many major pollutants. For instance, particulate matter can be blown from fields or roadways just as it can be generated by steel mills or cement plants. In such cases, the man-made share must first be assessed before required cutbacks can be determined.

15. For a comprehensive description and discussion, see Peter Bohm and Clifford S. Russell, "Comparative Analysis of Alternative Policy Instruments," in Allen V. Kneese and James L. Sweeney, eds., *Handbook of Natural Resource and Energy Economics*, vol. 1 (New York, North-Holland, 1985) pp. 395–460; see also Frederick R. Anderson, Allen V. Kneese, Phillip D. Reed, Serge Taylor, and Russell B. Stevenson, *Environmental Improvement Through Economic Incentives* (Washington, D.C., Resources for the Future, 1977).

16. For an interesting analysis of attitudes toward effluent charges, see Steven Kelman, *What Price Incentives?* (Boston, Auburn House, 1981).

17. William J. Baumol and Wallace E. Oates, "The Use of Standards and Prices for Protection of the Environment," *Swedish Journal of Economics* vol. 73 (March 1971) pp. 42–54.

18. These difficulties were quite serious, to be sure. Never before had the competence or commitment of the EPA's top management been questioned. Ultimately all but one of the agency's highest-ranking officials resigned or were fired. See Robert W. Crandall and Paul R. Portney, "Environmental Policy," in Paul R. Portney, ed., *Natural Resources and the Environment: The Reagan Approach* (Washington, D.C., Urban Institute, 1984) pp. 47–81.

19. Ibid., pp. 47–55.

three

Air Pollution Policy

Paul R. Portney*

For several reasons, 1981 promised to be the year in which significant changes would be made in the Clean Air Act, arguably the most important of all federal regulatory statutes. First, the act was due to be reauthorized that year. In past years reauthorization had proven to be a convenient opportunity to reconsider and amend that and other environmental laws. Second, President Reagan took office in January 1981 following a campaign in which he identified regulatory reform as one of the cornerstones of his economic recovery program.[1] When he quickly proposed sweeping tax and expenditure reductions to fulfill promises about the budgetary foundations of that program, it appeared as if the Clean Air Act might become a guinea pig that would demonstrate the kinds of changes his administration intended to make in federal regulations. And third, while there was little consensus about direction, there was an unusual amount of agitation for some sort of change in the Clean Air Act. Environmentalists, the business community, state and local officials, and

*For their very careful reading of, and constructive comments on, earlier drafts of this chapter, I wish to thank Robert Brenner, Robert W. Crandall, Maureen L. Cropper, Hadi Dowlatabadi, A. Myrick Freeman III, Michael Hazilla, Allen V. Kneese, Raymond J. Kopp, Alan J. Krupnick, Richard A. Liroff, Wallace E. Oates, William F. Pedersen, Jr., Phillip D. Reed, Clifford S. Russell, Robert M. Schwab, and V. Kerry Smith. The usual disclaimer applies.

other groups all had proposed modifications ready and waiting for the new administration and Congress.

It was not to be. Not until President Bush took office in January of 1989 was there any interest evinced in the White House in a serious review of the Clean Air Act. And in spite of the ambitious changes in the act proposed by President Bush in June 1989, passage of significant amendments is still far from certain. It is entirely possible that the 1970 and 1977 amendments to the Clean Air Act will constitute this nation's approach to air pollution control policy for some time to come.

The purpose of this chapter is to review and analyze the fundamentals of that approach. The first section presents a history and background of air pollution control efforts in the United States leading up to the major restructuring that took place in 1970. The second section outlines the major features of the new approach Congress adopted that year, emphasizing the substantial shift in responsibility from state and local governments to the federal government. The third section discusses several trends in air pollution and air quality since 1970. There care is taken to point out that these changes may not be exclusively the result of the 1970 legislation, a most important point to grasp. In the fourth section attention shifts to an economic evaluation of the current approach and the likely accomplishments of that approach. Evidence is presented on the costs and benefits of current air pollution control efforts (which bears on the wisdom of our air quality goals) as well as on the cost-effectiveness of the strategies chosen (are the goals being met as inexpensively as possible?). Where few or no data are available, the pros and cons of the Clean Air Act are discussed in qualitative terms. Some summary judgments are drawn and recommendations advanced in the concluding section.

AIR POLLUTION CONTROL BEFORE 1970

Neither air pollution nor efforts to control it are recent phenomena.[2] The earliest air pollution problems probably resulted from volcanic eruptions, the decomposition of organic matter, and other natural sources, and eventually from the first fires set by primitive people. While there are no doubt even earlier examples, one of the first modern pollution control ordinances can be traced to thirteenth-century England, where King Edward I banned the burning of certain highly polluting coals in London.[3]

Closer to home and nearer in time, air pollution was first combated in the United States through nuisance or trespass suits brought in the courts.[4] By the end of the Civil War, the industrial revolution had firmly taken hold. Not coincidentally, the very first air pollution statutes in the United States, designed to control smoke and soot from furnaces and locomotives, were passed by the cities of Chicago and Cincinnati in 1881. Within thirty years, county governments had begun to pass their own pollution control laws, and in 1952 Oregon became the first state to enact meaningful ordinances combating foul air. Other states followed, generally with legislation aimed at smoke and particulates; only California took on the air quality problems associated with motor vehicle exhausts. Table 3–1 shows the growth over time of city, county, and state laws protecting air quality.

In 1955 the federal government entered the picture for the first time with the passage of the Air Pollution Control Act. This law, which did little more than authorize federal funds to assist the states in air pollution research and in training technical and managerial personnel, was prompted by the states' agitation over dealing with what they thought was a national problem. The Air Pollution Control Act was extended by Congress in 1959 and again in 1962. Prior to both extensions, more ambitious powers were contemplated for the federal government but were scrapped in favor

Table 3–1. Development of U.S. Municipal, County, and State Air Pollution Control Legislation, 1880–1980

	Number of jurisdictions with statutes		
Year	Municipal[a]	County[b]	State
1880			
1890	2		
1900	5		
1910	23		
1920	40	1	
1930	51	2	
1940	52	3	
1950	80	2	
1960	84	17	8
1970	107	81	50
1980	81	142	50

Source: Adapted from Arthur C. Stern, "History of Air Pollution Legislation in the United States," *Journal of the Air Pollution Control Association* vol. 32, no. 1 (January 1982) p. 44.

[a] Includes city-county agencies.

[b] Includes multicounty agencies.

of merely continuing appropriations for research, technical assistance, and training for state air pollution control agencies.

The mid-1960s, however, saw the passage of three new acts that foreshadowed the later, more radical restructuring of federal air pollution control efforts. In 1963 Congress passed the original Clean Air Act. This law provided for permanent federal support for air pollution research, continued and increased federal assistance to the states for the development of their pollution control agencies, and, in an important departure, introduced a mechanism through which the federal government could assist the states where cross-boundary air pollution problems arose among them. In 1965 the Motor Vehicle Air Pollution Control Act was passed. Most important among its provisions, it permitted (but did not compel) the federal secretary of Health, Education, and Welfare (HEW) to establish emissions standards for new motor vehicles. This marked the real beginnings of an active federal role in controlling air pollution from mobile sources. Nevertheless, it stopped short of imposing actual vehicle emissions standards, something only the state of California had done by that time. However, in 1965 Congress also amended the Clean Air Act for the first time, directing HEW to set the first federal emissions standards for motor vehicles.

In 1967 Congress passed the Air Quality Act. This law—the temporal and philosophical precursor of the 1970 changes—again provided additional funds for the states, this time to help them plan as well as implement their pollution control strategies. More important, it required states to establish air quality control regions (now known as AQCRs), geographic areas that share common air quality concerns in much the same way so-called watersheds deal in common with water pollution problems. The act also directed HEW to investigate and publish information about the adverse health effects associated with a number of common air pollutants so that the states could then set air quality standards for them. Moreover, HEW (through its National Center for Air Pollution Control) was to identify viable pollution control techniques so that each state could get on with the business of regulating polluters to attain the air quality standards each was to have established.

This review of legislation up to 1970 is by necessity both brief and selective, but it conveys the overall evolution of federal policy before 1970. What began in 1955 with simple grants-in-aid to state and local governments gradually evolved to the point where those recipients were being given very specific responsibilities by the federal government to combat air pollution problems. The changes made in 1970 were in some sense the culmination of an increasing federal role.

THE NEW DIRECTION IN AIR POLLUTION CONTROL POLICY

Congressional impatience with the pace of activity under the pre-1970 approach, coupled with increasing public activism—exemplified by Earth Day in the spring of 1970—led inexorably to the Clean Air Act amendments of 1970 (the Clean Air Act as amended in 1970 is hereafter referred to simply as the Clean Air Act, or CAA). The perceived problems with the previous approach were several: HEW had been slow in issuing the guidance documents detailing the adverse health effects associated with common air pollutants; where these had been prepared, states had either failed to set air quality standards or were slow in developing implementation plans showing how they would meet the standards; and the automobile manufacturers, who had done little in the 1960s to inspire confidence in their commitment to pollution control, appeared capable of wriggling out of the emissions standards set by HEW.[5] These factors and others combined to produce a major overhaul in federal air pollution policy.

It is impossible to summarize neatly all the important features of the Clean Air Act. Indeed, the act as currently amended contains about one hundred very complicated sections and runs to 173 pages in length. Nevertheless, the basic structure is comprehensible. The simplest way to think of it is in terms of the two major kinds of responsibilities it creates, the first having to do with the goals of air pollution control and the second with the means of attaining them.

Goals

As a result of the Clean Air Act, the federal government—specifically, the administrator of the then newly created EPA—was charged with establishing the preeminent environmental objectives of federal air quality policy, the National Ambient Air Quality Standards (or NAAQSs). These standards, which were to be set by the states under the old approach, represent the maximum permissible concentrations of the common air pollutants which HEW had been studying at the time the 1970 amendments were made. Primary standards were to be those protecting human health; secondary standards were to be established if the health-based standards were insufficient to protect exposed materials, agricultural products, forests, or other non-health values.

Congress made several important decisions about the primary standards. First, they were to be uniform across the country. That is, each of the air quality control regions that had been established under the earlier legislation (or would subsequently be created) had to reduce ambient (or outdoor) air pollution concentrations at least to the level of the NAAQSs. The states could elect to impose stricter standards if they wished; however, no area could elect to have dirtier air than that called for in the NAAQSs. Second, Congress directed that the primary standards be set by the EPA at levels that would "provide an adequate margin of safety . . . to protect the public . . . from any known or anticipated adverse effects associated with such air pollutant[s] in the ambient air."

Both of these features are very important. National uniformity of standards means that areas where control costs are high or control benefits are low cannot respond by setting less stringent air quality standards to account for the situation; rather, such areas are locked into the same goals as other parts of the country. The margin-of-safety provision has become controversial because it embraces what is known as the "threshold model" of air pollution-related illness. That is, by calling for a margin of safety in primary standard-setting, Congress implicitly signaled its belief that "safe" levels existed and could be identified for the common air pollutants. Once this was done for a particular pollutant, each NAAQS was to be set at a level below that threshold, thus providing the putative margin of safety.

This concept is illustrated in figure 3–1, where air pollution concentrations are graphed against excess sick days (above some natural background of occasional illness), a hypothetical measure of the health of the population. Line segment AA′, illustrating one possible relationship, indicates that for pollution concentrations between O and A, no excess adverse health effects occur. Beyond point A—the threshold concentration—increases in pollution imply more illness. Thus, according to the threshold approach, the NAAQS for this hypothetical pollutant might be set at OB, providing a margin of safety equal to BA.

Two problems have arisen with this approach. The first arises because a growing body of physiological and toxicological evidence suggests that there may be *no* safe levels for any of the air pollutants for which NAAQSs must be set. That is, there may be individuals so sensitive to pollution that virtually any positive concentration may increase their risk of illness or discomfort.[6] If so, the true dose-response curve linking pollution to illness may look more like the curve OC. While flat at low ranges, this curve nevertheless indicates that even very low pollution concentrations may result in some excess risk of illness. The following problem must then be faced: if no threshold exists, the only concentration providing the statutorily required margin of safety is zero. Yet setting permissible concentrations at zero would imply an end to all fossil fuel combustion and to commercial and industrial activity—indeed, to modern society as we

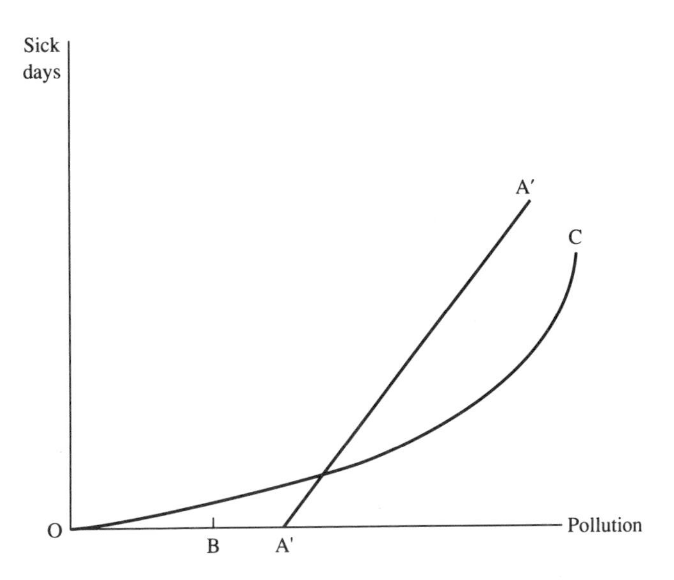

Figure 3–1. Dose-response functions linking air pollution to illness

know it.[7] Since Congress clearly did not contemplate such extremes, the EPA administrator is caught between the apparent mandate of the law and the realities of science and economics.

The second problem with the standard-setting process is more practical. Under current interpretations, the NAAQSs are to be set without regard to the costs of attainment (much like the zero-risk approach discussed in chapter 2). Yet even if threshold concentrations could be identified, might not the costs of meeting standards set at those levels (as opposed, say, to slightly less protective standards) be so great that the latter standards would be preferred? Even if this were so such a decision appears to be prohibited under the Clean Air Act.

The importance of the NAAQSs cannot be overemphasized. They are the goals to which a substantial part of the air pollution control activity in the United States is geared. With the possible exception of the emissions standards for all new sources (discussed in the next subsection), these ambient standards are the most important part of the Clean Air Act as revised in 1970. Table 3–2 presents the NAAQSs for the six common air pollutants as they stood in 1989.

Congress also expressed concern in 1970 with less common air pollutants that might pose serious threats to health. Rather than call for maximum

Table 3–2. National Ambient Air Quality Standards (NAAQSs) in Effect in 1989

Pollutant	Primary (health related)		Secondary (welfare related)	
	Averaging time	Standard level concentration[a]	Averaging time	Concentration
Particulate matter (PM_{10})[b]	Annual arithmetic mean	50 $\mu g/m^3$	Same as primary	
	24-hour	150 $\mu g/m^3$	Same as primary	
Sulfur dioxide (SO_2)	Annual arithmetic mean	(0.03 ppm) 80 $\mu g/m^3$	3-hour	1300 $\mu g/m^3$ (0.50 ppm)
	24-hour	(0.14 ppm) 365 $\mu g/m^3$		
Carbon monoxide (CO)	8-hour	9 ppm (10 $\mu g/m^3$)	No secondary standard	
	1-hour	35 ppm (40 $\mu g/m^3$)	No secondary standard	

Nitrogen oxide (NO_2)	Annual arithmetic mean	0.053 ppm (100 $\mu g/m^3$)	Same as primary
Ozone (O_3)	Maximum daily 1-hour average	0.12 ppm[c] (235 $\mu g/m^3$)	Same as primary
Lead (Pb)	Maximum quarterly average	1.5 $\mu g/m^3$	Same as primary

Source: Office of Planning and Standards, Environmental Protection Agency, *National Air Quality and Emissions Trend Report, 1987* (Research Triangle Park, N.C., March 1989) p. 22.

[a] Parenthetical value is an approximately equivalent concentration.

[b] New PM standards were promulgated in 1987, using PM_{10} (particles less than 10μ in diameter) as the new indicator pollutant. The 24-hour standard is attained when the expected number of days per calendar year above 150 $\mu g/m^3$ is equal to or less than 1, as determined in accordance with Appendix K of the PM NAAQS.

[c] The standard is attained when the expected number of days per calendar year with maximum hourly average concentrations above 0.12 ppm is equal to or less than 1, as determined in accordance with Appendix H of the ozone NAAQS.

ambient concentrations to be set, as with the NAAQSs, Congress directed the EPA administrator to identify any such pollutants (now referred to as toxic or hazardous air pollutants) and propose discharge standards for the major sources that emit them. However, in wording quite similar to that concerning the common air pollutants, the EPA administrator was directed by Congress to set these emissions standards so that the concentrations remaining in the air after controls were applied would be so low as to provide "an ample margin of safety" against adverse health effects. In essense, then, a similar environmental goal was established for both common and hazardous air pollutants.

What about those parts of the country where air quality was already better than the national air quality standards established by the EPA? Would new industrial and other growth be allowed there until air quality had deteriorated to the level of the national standards? While the 1970 amendments were virtually silent on this subject, a landmark court case soon established that the Clean Air Act was to be implemented in such a way as to preserve air quality in clean regions as well as enhance it in polluted areas.[8] In 1977 Congress amended the Clean Air Act once again to formally declare the "prevention of significant deterioration" (PSD) in clean areas to be an additional goal of the act. The 1977 amendments established three classes of already clean areas. In the Class I areas (which include national parks and similar federal areas, as well as any other areas a state elects to include), very little additional deterioration in air quality is permitted, even if current concentrations are far below the NAAQS. Somewhat more pollution is permitted in Class II areas (which comprise most of the rest of the clean air regions), while in Class III areas air quality is permitted to deteriorate up to but not beyond the level of the NAAQS.

The same 1977 amendments established another goal of air pollution control policy—the protection and enhancement of visibility in the national parks and wilderness areas of the United States. Thus, even where air pollution was not threatening health, it was to be reduced if it impaired, say, a visitor's view of the Grand Canyon or the Yosemite Valley.

These then were the environmental goals Congress established through the Clean Air Act as amended in 1970 and 1977. As we turn our attention to the means of attaining them, it will become clear that the adoption of advanced technologies was another goal of the act, quite apart from its relation to specific environmental standards.

Emissions Standards as a Means of Attainment

While Congress gave the federal government the sole responsibility for goal-setting under the Clean Air Act, the authority for accomplishing these goals was more evenly divided. The federal government appears to have been given the upper hand here, too; nevertheless, as we will see, some very important powers were reserved for lower levels of government.

Mobile sources. On the unassailable logic that fifty different sets of state standards would play havoc with motor vehicle manufacturers, and to spur carmakers into serious action on pollution control, the establishment of emissions standards for cars, trucks, and buses was to be handled in Washington.[9] In fact, Congress took the unusual step of writing into the Clean Air Act itself the emissions reductions that cars (and later trucks) would have to meet, as well as the schedule for their accomplishment. (In contrast, in all but one other part of the act the issuance of specific, numerical discharge standards was delegated to the EPA or the states.) The vehicle standards Congress adopted called for a 90 percent reduction in average hydrocarbon and carbon monoxide emissions by 1975 (measured from the already controlled levels existing at the time) and the equivalent of an 82 percent reduction in the nitrogen oxides emitted by cars (measured from then-uncontrolled levels). The EPA was empowered to waive the deadlines for meeting these standards under certain circumstances (which it has done regularly), and was also given the power to issue emissions standards for buses and motorcycles.

Several features of these emissions standards are worth noting because they are common to other emissions controls mandated under the CAA. For one thing, the congressionally mandated pollution standards for motor vehicles were not connected in any clear way to the environmental goals of the act as expressed through the NAAQSs; that is, the vehicle standards applied equally to cars in very dirty and very clean areas. For another, responsibility for reducing auto pollution was laid almost entirely on the shoulders of vehicle manufacturers—until the recent advent of state vehicle inspection programs, little was required of vehicle owners.

Stationary sources. What about the controls on factories, power plants, refineries, and other industrial facilities, known collectively as stationary sources? Were emissions limits on these sources intended to be keyed to local air quality, or—like the controls on cars and other vehicles—to be uniform across the country? A little bit of both, as it turns out.

Without a doubt, the most important provision introduced in the Clean Air Act relating to stationary sources was that pertaining to newly constructed (or substantially modified) plants. In section 111 of the act, Congress gave the Environmental Protection Agency the power to set binding emissions standards for all new sources of the common air pollutants. These are known as new source performance standards (NSPSs) and have two important characteristics: first, the EPA must base

NSPSs on the "best technological system of continuous emission reduction" (in other words, these are to be technology-based standards, determined by the state-of-the-art in pollution control at the time the standards are set); and second, the new source standards must be affordable by affected parties. The latter requirement does not introduce balancing into the NSPS process, since the controls are assumed to be worthwhile so long as the source can afford them. But the requirement does introduce—at least crudely—some economic considerations into the standard-setting process. The NSPSs were initially to apply uniformly throughout the United States, irrespective of air quality in the area where a plant was being built.

Over time, some variation based on local conditions has been introduced into the new source emissions standards. Congress decided that in areas where at least one of the NAAQSs was being violated (called nonattainment areas), new sources had to meet even stricter emissions standards than those called for by the new source performance standards. In 1977 Congress amended the Clean Air Act to require new sources wishing to locate in nonattainment areas to install technology consistent with the "lowest achievable emissions rate" (now known as LAER).[10] Furthermore, in most PSD areas (where air quality is even better than that called for by the NAAQSs), major new sources were required to install the "best available control technologies" (or BACT).

Readers confused by the distinctions between NSPS, LAER, and BACT should not despair of their own analytic capabilities. Defining and distinguishing between these levels of technological control for new sources have tied the Environmental Protection Agency, the courts, and the regulated community in knots. As a practical matter, however, if the EPA has defined a new source performance standard for a particular kind of new source, that standard is often considered satisfactory as LAER or BACT in nonattainment or PSD areas, respectively, even though LAER and BACT are in principle supposed to be more strict than NSPS emissions limits.

With the Environmental Protection Agency authorized to establish more or less uniform emissions standards for all new plants, the only unfinished business in the Clean Air Act concerned the controls to be imposed on *existing* sources in operation in 1970. Here, in an important departure from the overall shift to federal authority, the states were given the responsibility for establishing limits. Specifically, the Clean Air Act mandated that each state prepare a state implementation plan (or SIP) that would demonstrate how existing sources would be controlled. All parts of the country were to be in compliance with the NAAQSs by 1975 (later extended to 1977, 1982, 1987, and most recently 1988, at least for some pollutants). The EPA did retain the authority to reject a state's plan if it was inadequate to assure attainment and, in the limit, to step in and impose federal controls to meet the NAAQSs.

This division of responsibility gave the states and local governments primary control over existing sources, which was important to them from a political standpoint. When the federal EPA imposes strict standards for all new plants to meet, the construction of new plants may be slowed and opportunities for new jobs may be reduced. But the losers from such a policy are difficult to identify (since it cannot be determined just who would have gotten the new jobs had a plant been built) and hence politically weak. On the other hand, stringent controls imposed on existing plants can result in their closure and thus endanger the jobs of current workers. Because this could bring considerable political pressure to bear on local officials, they wanted—and got—the responsibility for regulating existing local sources.

Local control of existing local sources has had another important effect. Suppose pollution in one jurisdiction could be transported by prevailing winds to another jurisdiction (as we know to be the case). In such an instance, the first (or pollution-exporting) region could impose weak controls on its local sources, thereby enhancing its economic position, with little fear of suffering from the resulting pollution. On the other hand, the area receiving the pollution might be controlling the sources within its jurisdiction quite stringently, only to suffer from imported pollution. While the CAA includes language which, in principle, gives the EPA the power to act in such cases, in practice it has proved to be inadequate. The problem of acid rain has arisen in part because of the principle of local control of existing sources.

One final important responsibility was reserved for state and local governments—that of monitoring for and enforcing compliance with both the environmental goals and the individual source discharge standards called for in the Clean Air Act. This provision was in part a simple continuation of the pre-1970 policy under which localities began to monitor air pollutants, particularly sulfur dioxide and total suspended particulate matter; consistent with that continuation, monies were appropriated under the CAA for continued financial assistance to local air pollution agencies. The policy of local monitoring and enforcement also served a political purpose: it gave local areas the flexibility to coax local sources into compliance. In practice, some recalcitrants have thumbed their noses at local enforcement efforts; for this reason, the federal EPA can and has from time to time entered the picture with back-up enforcement authority.

This, then, is the structure that emerged in the Clean Air Act. The federal government sets the air quality goals that all parts of the country are to meet but also limits degradation in the already clean areas.

Emissions standards for motor vehicles, for all new stationary sources of the common air pollutants, and for sources of toxic air pollutants are also to be set in Washington (by Congress in the case of cars, and by the EPA in all other cases). Against the backdrop of these new source standards, individual states are to impose appropriate limits on existing sources to ensure that the national air quality standards are met. Finally, the states are to monitor and enforce the whole system of emissions limitations.

ACCOMPLISHMENTS SINCE 1970

One of the things we would most like to know about the 1970 redirection in air pollution policy is the effect it has had on air quality. For a number of reasons, however, this is more difficult to determine than it might seem. First, air quality can be described in a variety of ways—for instance, by average pollutant concentrations, frequency of violations of the NAAQSs, extremely hazardous incidents, and so on. While these measures will generally move together, it is possible for improvements to be made along one dimension but not in another. Second, it is a bit heroic to think of air quality in national terms. Inevitably, some areas will improve while others stay the same or worsen, regardless of the measures used. Such geographic variation is lost with national averages. Third, the air quality monitoring network upon which a trends analysis must be based is simply inadequate. Although this deficiency is not discussed in detail in this chapter, it is hard to resist pointing out that a country spending as much for air pollution control as the United States should be ashamed of the thinness of its monitoring efforts.[11] Even absent these problems, it is extremely difficult to isolate the effects of regulatory policies on air quality, as distinct from the effects of other potentially important factors (we return to this point momentarily).

In spite of these difficulties, however, there are some data worth reviewing.

Ambient Air Quality

According to the Environmental Protection Agency, air quality data before the mid-1970s are of questionable quality for establishing national trends.[12] For that reason, the best national trends data begin in the mid-1970s. Trends for the six pollutants for which NAAQSs were in effect until 1987 are illustrated in figure 3–2: panels (a) through (f) are for total suspended particulates (TSP), sulfur dioxide (SO_2), carbon monoxide (CO), nitrogen dioxide (NO_2), ozone (O_3), and lead (Pb), respectively.

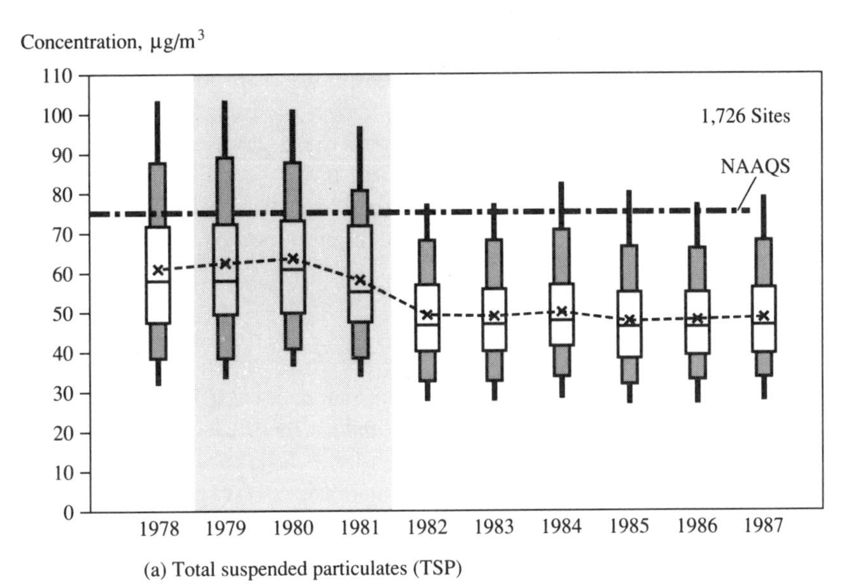

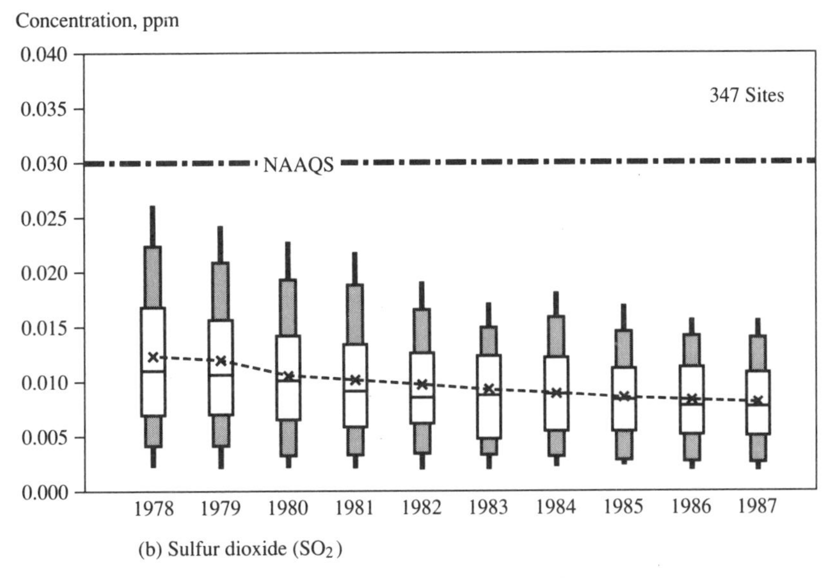

Figure 3–2. National air quality trends, 1978–1987. Standing in panels (a) and (e) indicates years during which the EPA was less confident in monitoring measurement. *Source:* Office of Air Quality Planning and Standards, Environmental Protection Agency, *National Air Quality and Emissions Trends Report, 1987* (Research Triangle Park, N.C., March 1989).

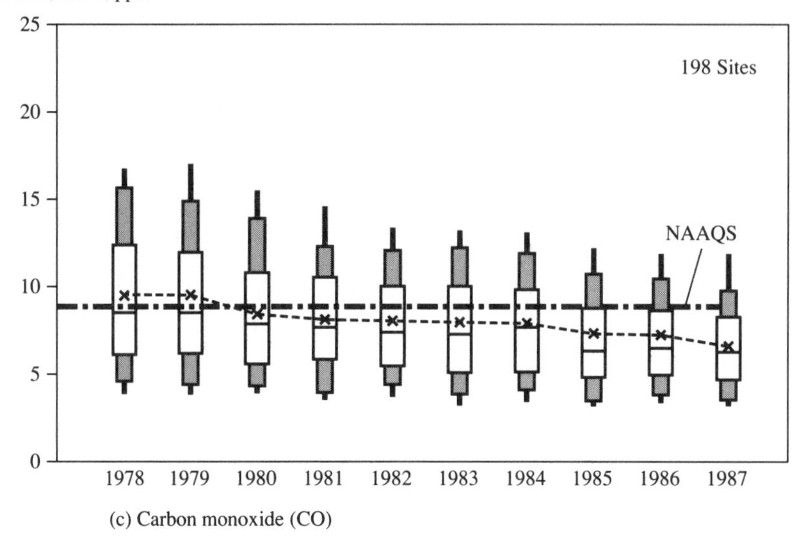

(c) Carbon monoxide (CO)

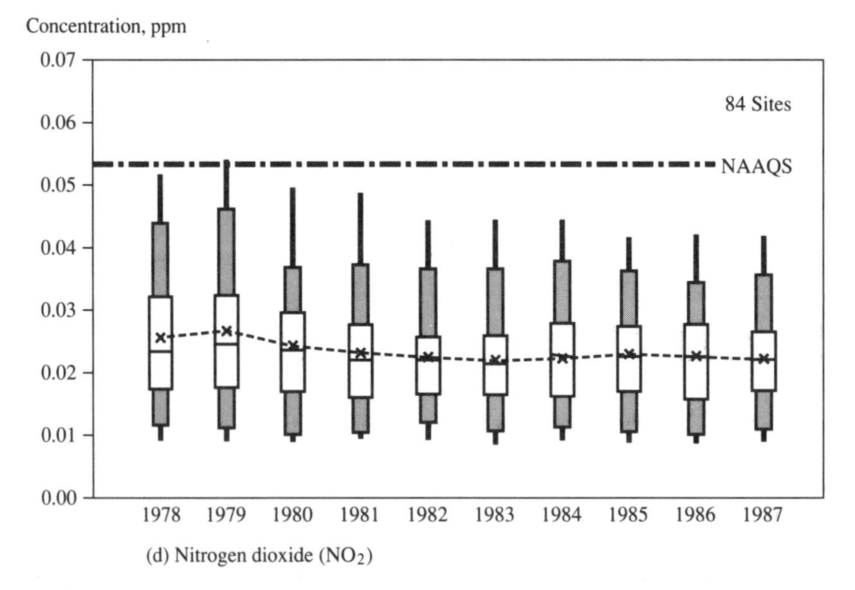

(d) Nitrogen dioxide (NO_2)

Figure 3–2. National air quality trends, 1978–1987 (*continued*)

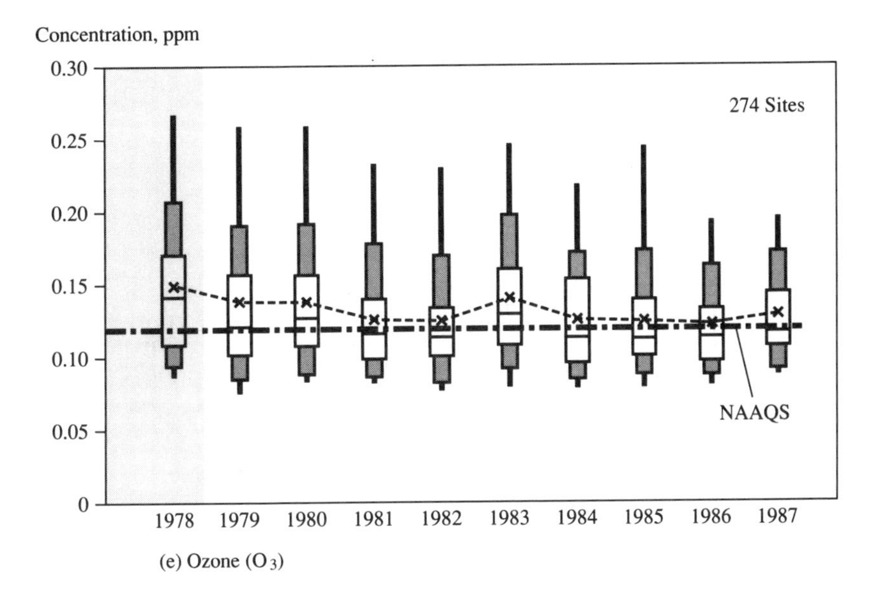

(e) Ozone (O_3)

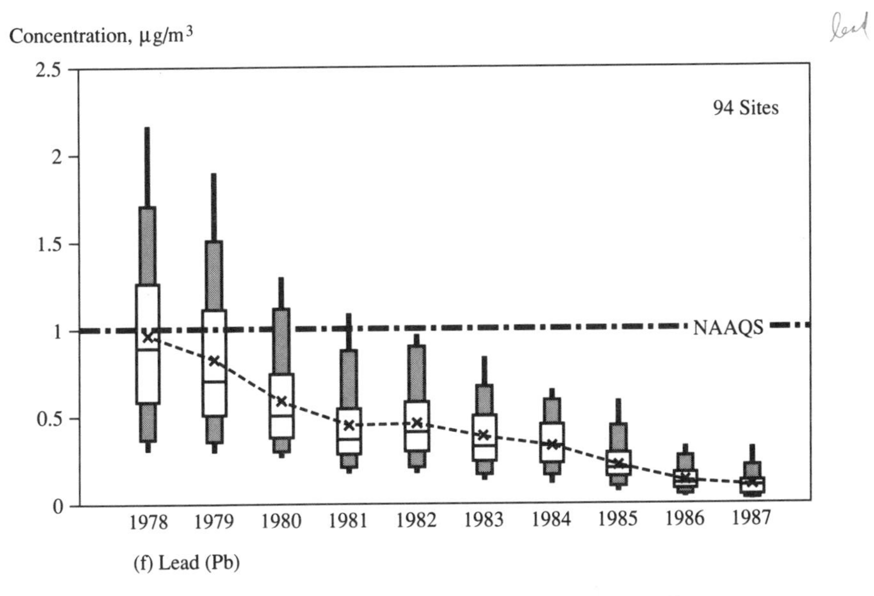

(f) Lead (Pb)

Figure 3–2. National air quality trends, 1978–1987 (*continued*)

Taken as a whole, these national trends appear encouraging. Annual average TSP concentrations (as measured at 1,726 sites around the country) fell about 21 percent between 1978 and 1987. Annual average concentrations of SO_2 (averaged over 347 sites) fell even more over the same period, by more than 35 percent. Similarly, annual ambient concentrations of CO fell by 32 percent between 1978 and 1987 at 198 U.S. monitoring sites. Lead concentrations fell even more precipitously—nearly 88 percent over the same period (based on data from just 97 urban sites). Finally, smaller improvements were recorded for NO_2 (12 percent, 84 sites) and O_3 (9 percent, 274 sites). Note that with the exception of ozone, the *average* concentrations across the sites (represented by the *x* in the middle of the box plot) are well below the relevant NAAQSs, indicated by the solid horizontal lines. If we are willing to overlook the sometimes questionable data and the relatively small number of monitoring sites, the overall trend in air quality appears to be a favorable one.

Attainment status. In spite of the fairly steady reductions in *average* ambient concentrations, the NAAQSs are still violated much more often than is permitted under the Clean Air Act. The problem of widespread nonattainment, as it has come to be called, is one of the most vexing in current air pollution policy.

Under the Clean Air Act, no more than one violation per year is permitted in an AQCR for each of the one-hour, eight-hour, or twenty-four hour standards (technically, the one-hour ozone standard cannot be violated more than three times during any three-year period). In actuality, however, violations are quite common in many metropolitan areas. For instance, although violations in Los Angeles have fallen significantly, the ozone standard there was exceeded an average of 123 times each year between 1983 and 1985.[13] It is the ozone standard that is most often violated in the United States; according to the EPA, at the time of writing nearly 370 counties or equivalent areas (with a combined population of 95 million persons) witness two or more days per year during which the one-hour ozone standard is exceeded. Table 3–3 shows the extent of nonattainment for ozone and other criteria air pollutants as of the end of 1988.

Beyond the frequency of violations, the seriousness of the nonattainment problem is more difficult to assess. As is too often the case, more data are needed. Among the questions needing answers are the following: In areas where persistent violations of one or more of the NAAQSs occur, how serious are those violations? (In Los Angeles, for instance, ambient ozone concentrations are sometimes triple the maximum permissible level of 0.12 ppm; elsewhere the standards are often barely exceeded, thus

Table 3-3. Nonattainment Areas in Counties or Parts of Counties, January 1989

Criteria air pollutant	No. of counties or parts of counties having nonattainment areas
Carbon monoxide (CO)	123
Nitrogen dioxide (NO_2)	4
Ozone (O_3)	341
Sulfur dioxide (SO_2)	67
Total suspended particulates (TSP)	317
Total nonattainment areas	852
Total no. of U.S. counties	529

Source: Information supplied by the Office of Air Quality Planning and Standards, Environmental Protection Agency, June 27, 1989.

giving less cause for concern.) In areas where multiple violations are the rule, do they all occur at the same monitoring station(s) or do they occur more randomly? (Any day on which the reading at *any* one of the monitoring stations exceeds the standard is counted as a day of violation; thus, if each of ten monitors in a metropolitan area exceeds the relevant ambient standard only once per year but on different days, that area will be classified as having ten yearly violations.) Presumably we would be more concerned about persistent violations at any one monitor, especially if it is located in a heavily populated area.

The most difficult question about the nonattainment problem is one we cannot attempt to answer here: What improvements in human health would result from eliminating most or all violations of the NAAQSs? Unfortunately, no unambiguous answer to this question is possible given any amount of space, owing to the great uncertainty and conflicting evidence about the effects of any of the criteria air pollutants on human health.[14]

Nevertheless, several generalizations seem warranted. The fairly steady improvements in urban air quality described above have eliminated many of the more immediate or acute threats to health associated with air pollution. It is highly unlikely, for instance, that the United States will ever again experience episodes involving the criteria pollutants serious enough to trigger significant premature mortality like that which occurred in Donora, Pennsylvania in 1948, when a temperature inversion trapped a thick blanket of smoke over the city for an extended period. Yet ambient concentrations are clearly high enough in some areas during certain periods that even healthy individuals experience considerable discomfort and acute adverse health effects (smog in the Los Angeles basin being a definitive example). Of the pollutants that pose the most acute threats to health *at existing concentrations*, ozone and sulfates—particularly acidic sulfates—appear to be the most harmful.

While existing air pollution concentrations may be sufficient to cause or at least exacerbate chronic respiratory or other illnesses, it is difficult to establish the likelihood or magnitude of such effects.[15] In a sense we are victims of our own successes here: pollution levels are low enough in most places now that their effects on chronic illness are often obscured by other causes, foremost among these being smoking (including passive smoking) and certain occupational exposures to harmful substances. However, the reduction in ambient lead concentrations over the past decade will almost surely have significantly favorable effects on chronic illnesses.

International experience. For purposes of comparison, figure 3–3 presents data on ambient concentrations of suspended particulate matter from a number of cities around the world that participate in a United Nations monitoring program. It can be seen that for the period 1976 to 1980, at least, urban air quality was quite variable worldwide. In cities like Tokyo and New York, particulate concentrations averaged about 60–70 $\mu g/m^3$, while they were nearly ten times that in Tehran and Calcutta (note the logarithmic scale on the vertical axis). It is important to remember that cities in the United States that are relatively polluted by our standards might be considered quite clean in other parts of the world.

Pollutant Emissions

Many discussions of air pollution policy are based on data on pollution emissions rather than ambient concentrations. In general, reliance on the former is to be discouraged because it shifts attention away from environmental quality, which ought to be the focus of environmental policy. Occasionally, however, emissions data can be helpful. Suppose that many industrial plants previously located in urban areas (where air pollution is monitored) are shut down and new plants are opened up in unmonitored rural areas. Even if total pollutant emissions stay the same, monitoring data would show a sharp drop in ambient concentrations that would be at least somewhat misleading.[16] Also, some pollutants (SO_2, for

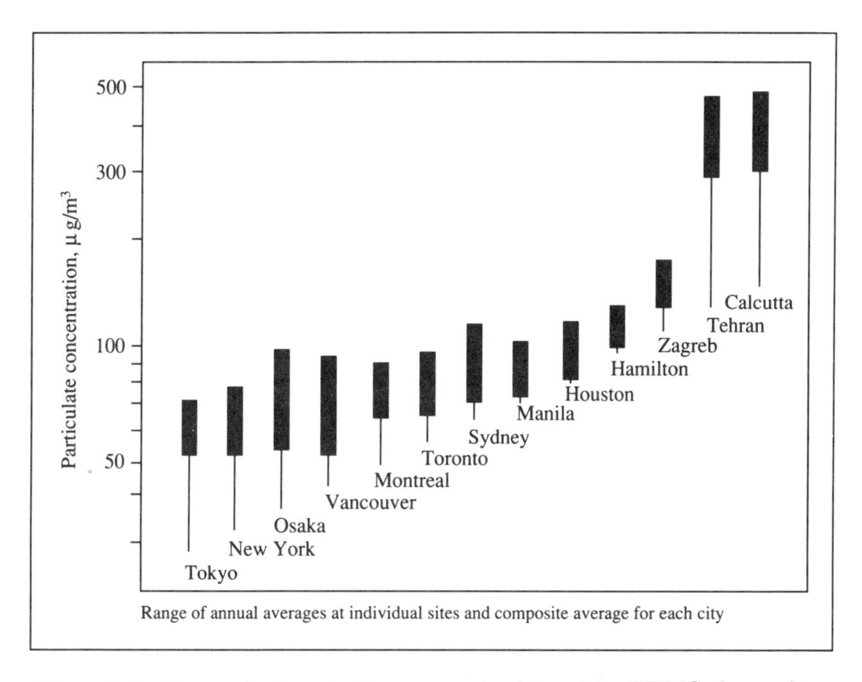

Figure 3–3. Suspended particulate matter in cities of the GEMS air monitoring network, 1976–1980. GEMS = Global Environmental Monitoring System. *Source:* Burton Bennett, Jan Kretzchmar, Gerald Akland, and Henk de Koning, "Urban Air Pollution Worldwide," *Environmental Science and Technology* vol. 19, no. 4 (1985) p. 303.

instance) can undergo chemical transformation after they are emitted; rather than remain suspended in ambient air, they may combine with water vapor or attach to airborne particles and return to earth in wet or dry form—sometimes quite acidic. The more likely such a transformation is, the less useful are measures of ambient concentrations as a guide to policy. Here too emissions data may be useful.

Table 3–4 presents data on estimated pollutant emissions in the United States between 1940 and 1987. According to these EPA estimates, emissions of all the common pollutants except nitrogen dioxide fell between 1970 and 1987. Moreover, the percentage reductions were substantial, particularly for lead, estimated emissions of which were reduced by 96 percent (primarily through the required use of lead-free gasoline in all new cars). Estimated particulate emissions were reduced

Table 3–4. Summary of Estimates of Nationwide Emissions, 1940–1987

Pollutant (10^6 short tons/year)	1940	1950	1960	1970	1975	1980	1981	1982	1983	1984	1985	1986	1987
Particulate matter (PM/TSP)	25.5	27.4	23.8	20.4	11.6	9.4	8.8	7.8	7.8	8.1	7.8	7.4	7.7
Sulfur oxides	19.4	21.8	21.7	31.2	28.5	25.8	24.9	23.6	22.8	23.7	23.3	22.8	22.5
Nitrogen oxides	7.5	10.3	14.1	20.1	21.1	22.5	22.5	21.6	20.9	21.7	21.8	21.2	21.4
Reactive volatile organic compounds	20.0	22.3	24.9	28.9	24.3	24.6	23.1	21.7	22.5	23.7	22.1	21.3	21.6
Carbon monoxide	89.8	94.9	97.1	110.4	90.6	84.8	82.0	76.4	78.6	75.8	71.2	67.4	67.7
Lead (10^3 short tons/year)	NA	NA	NA	224.6	162.1	77.8	61.7	60.0	51.1	44.2	23.3	9.5	8.9

Pollutant	% Change 1940–1987	% Change 1970–1987	% Change 1980–1987	% Change 1986–1987
Particulate matter (PM/TSP)	−70	−62	−18	4
Sulfur oxides	16	−28	−13	−1
Nitrogen oxides	186	7	−5	1
Reactive volatile organic compounds	8	−25	−12	1
Carbon monoxide	−25	−39	−20	0
Lead	NA	−96	−89	−6

Note: A value of zero indicates emissions of less than 50,000 metric tons. NA = not available.

Source: Environmental Protection Agency, "National Emissions Estimates, 1940–1987," Document no. EPA-450/4-88-022 (March 1989) p. 2.

more than 60 percent, while those for sulfur dioxide, carbon monoxide, and volatile organic compounds (a precursor of ozone) fell by 28, 39, and 25 percent, respectively. Emissions of nitrogen dioxide are estimated to have increased by 7 percent between 1970 and 1987.

Linking Policies to Accomplishments

If the data on ambient concentrations and estimated emissions can be taken at face value, cannot one conclude that the Clean Air Act has been successful? After all, both appear to have fallen since the act has been in effect. The answer is, not necessarily, as the following examples may demonstrate. Suppose—hypothetically—that simultaneously all U.S. pollution control requirements were removed and the U.S. and world economies entered a severe recession. Pollutant emissions might fall dramatically (and with them ambient concentrations) because the production of autos, steel, chemicals, gasoline, and other products of "dirty" industries were being curtailed. Yet no one would argue that the lifting of pollution controls caused air quality to improve. Obviously, other factors were at work. To take another example, a short-term or even seasonal temperature inversion in a particular area might trap pollutants and cause ambient concentrations to increase even though pollution controls were actually reducing emissions.

The point is that both pollutant emissions and ambient concentrations depend on a number of factors. These include environmental regulations, to be sure. But they also include other, nonregulatory factors like the general level of business activity; the rate and type of economic growth in particular areas; the prices of coal, oil, wood, gasoline, and other fuels used in home furnaces, cars, and industrial boilers; and even the weather.[17] Before passing judgment on the efficacy of the CAA, these other factors must somehow be netted out. This is no easy task—although the isolation of variables is a common problem in the social sciences—since controlled experiments in which these other factors can be held constant generally are not possible.

Nevertheless, researchers using modern statistical methods have recently analyzed changes in both ambient air pollution concentrations and industrial pollution emissions in an attempt to isolate the effects of air quality regulations from other factors.[18] Both studies had difficulty identifying any regulatory effect. The first examined changes in TSP concentrations between 1973 and 1977 in a sample of 93 metropolitan areas. This study included variables measuring the weather, the level and composition of industrial activity, and the pollutant content in the fuels burned by households and electric utilities. In general, no correlation was found between either pollution control investment or the pollutant content of fuels on the one hand and air quality on the other, even though the former two are at least crude measures of regulation. An inverse association between TSP levels and pollution control investment was found when the analysis was restricted to the eastern United States, however.

The second study attempted to relate total pollutant emissions by industry to each industry's investments in pollution control, annual coal usage, output price, and total investment in plant and equipment.[19] Overall, the study concluded that investment in pollution control plays only a limited role in explaining pollutant emissions by industry, and that economic variables like coal usage, total investment, and product price have much greater explanatory power.

Both of these studies are important because they attempt, for the first time, to pinpoint the effects of air quality regulation while holding constant other important influences on emissions and environmental quality. However, both also suffer from shortcomings that limit the confidence we can place in them. Through no fault of the investigators, for example, both studies were based on data of quite questionable value. As pointed out above, emissions data are virtually all estimated by the EPA rather than actually monitored. Thus the dependent variable (that which is being explained) in the second study may have been measured with systematic error. The data on pollution control expenditures by industry or by metropolitan area are also of questionable value: they are generally extrapolated from a sample of firms or plants and can diverge considerably depending on who asks the questions.[20] Data at the level of individual firms or plants are almost essential if this kind of analysis is to achieve its potential.

The Record Before 1970

A final piece of evidence to consider in thinking about the Clean Air Act and its possible effects is the trend in air quality before the act. If air quality had been deteriorating prior to 1970 but then began to improve, some contribution on the part of the amendments passed that year might be suggested. While the air quality data extending back into the 1960s are less reliable than today's, they do tell an interesting story.

According to data from the EPA, average ambient TSP levels fell about 22 percent between 1960 and 1970 (as monitored at 95 sites around the United States).[21] During the period 1966 to 1971 annual average ambient SO_2 concentrations fell by an even larger 50 percent (based on data from 31 sites). While we must be leery of trends based on such a small number of sites, these data are important because they suggest that air quality was improving as fast or faster before the Clean Air Act as it has since that time. Like the two studies cited above, this conclusion should give us pause in reflecting on the likely effects of the CAA.

These data also call into question one of the fundamental premises behind the act—that states and local governments would never impose the controls necessary to achieve healthful air. While some of the pre-1970 improvements were no doubt due to economic factors of the kind discussed above, they were also hastened by state and local ordinances regulating the incineration of garbage and the burning of coal or high-sulfur fuel oil in residential, commercial, and industrial furnaces. It is arguable whether local governments acting alone could have continued to make progress in the fight against air pollution; certainly it would have been difficult for them to deal with those pollutants emanating from smaller and more diffuse sources. Nevertheless, their accomplishments prior to 1970 should not be ignored.

Summary

Despite somewhat conflicting evidence, the following conclusions about the U.S. air quality experience since 1970 seem warranted. To begin with, air quality in the United States appears to have improved in most places since 1970. True, part of this improvement may be due to the closure of some large sources of pollutants or their relocation to remote, unmonitored regions. It is also true that some metropolitan areas—particularly in southern California—are still experiencing far too many days on which some NAAQSs are being violated, while other areas have seen air quality deteriorate. Nevertheless, most people in large metropolitan areas have experienced improved air quality over the last decade.

Total emissions of the common air pollutants (with the exception of nitrogen dioxide) also appear to have declined, although this conclusion is colored by the unfortunate practice of estimating rather than monitoring actual emissions from most sources. This suggests that the gains in ambient air quality are not due to source relocation alone.

Linking these changes to the Clean Air Act is more problematic. Other factors (the economy, fuel prices, the weather, for example) also play a role in explaining emissions and/or ambient concentrations. When these factors have been included in analyses of emissions or air quality changes, the role of regulation seems small, although such analyses are preliminary and limited in important ways. The fact that at least some measures of air quality were improving at an impressive rate before 1970 suggests that other factors in addition to the CAA are behind the recent improvements.

Nevertheless, it seems indisputable that the Clean Air Act has played an important role in improving U.S. air quality or in preventing its further degradation. For this not to be so, one would have to believe that all the investments in both stationary and mobile source pollution control have been for naught. Yet tons and tons of scrubber sludge, flyash, and other solid wastes attest to the partial efficacy—at least—of air pollution control equipment on factories and power plants. Similarly, while there can be no disputing the deterioration, over time, of catalytic converters and other automobile pollution controls, it seems clear that the CAA provisions dealing with mobile sources have had a positive impact on air quality.[22]

Even where air quality has deteriorated since 1970, it is probably fair to say that the deterioration would have been worse had it not been for the controls imposed by the CAA on stationary, and especially mobile, sources. Ultimately, of course, this is the true measure of any regulatory or other government intervention: how different does the world look *with* the program in question from how it would have looked *without* it? Since other things generally do change, this is a different question than asking what has happened since the program took effect.

A final point remains to be made. While the above judgment about the role of the Clean Air Act in improving air quality may be comforting, much more needs to be known before the act can be declared a success or a failure. This ultimate judgment necessitates comparing the improvements in air quality associated with the act (and the resulting enhancement of human welfare) with the sacrifices society has been forced to make to obtain them. In other words, we must compare the benefits and costs associated with the CAA. It is to this and related matters that we now turn.

AN ECONOMIC EVALUATION OF THE CLEAN AIR ACT

Before plunging directly into estimates of the benefits and costs associated with the Clean Air Act, it is useful to explain briefly how such estimates are often made. (More comprehensive treatments are available elsewhere for the interested reader.)[23] After subsequently reviewing some of the quantitative evidence on benefits and costs, we examine certain features of the CAA which are of economic significance but about which we have little hard evidence.

Fundamentals of Benefit and Cost Estimation

Comprehensive benefit estimation under the Clean Air Act can be thought of as a four-step process. First, regulations written pursuant to the

Clean Air Act must be translated into reduced emissions by the affected sources, both stationary and mobile. Next, reduced emissions must be converted into reduced ambient concentrations of TSP, SO_2, O_3, and other regulated pollutants. Then the reduced ambient concentrations must be mapped into, or related to, enhanced human welfare or well-being (which may take such forms as improved health, increased visibility, or reduced damage to crops, forests, and exposed materials). Finally, individuals' *willingness to pay* for these gains must somehow be ascertained, since that is how benefits are measured in dollar terms in formal benefit-cost analysis.

Even at this level of generality, one can imagine the difficult problems that crop up at each stage in estimating benefits. For instance, the magnitude of actual emission reductions resulting from a given technology-based regulation will depend upon the efficacy of the control technology selected, the effort sources make to operate and maintain equipment properly, and natural forces beyond the control of the sources. Similarly, ambient environmental conditions depend not only on emissions control, but also on meteorological, topographical, and other factors as well, as we saw earlier. The third step—determining the effects of changes in ambient conditions on individuals, ecosystems, and materials— may be the most difficult of all. An individual's health, for example, depends in important ways on factors other than the environment, like cigarette smoking, genetic disposition, income, and education. The weather, certain types of pests, and many other factors can affect natural ecosystems or exposed materials. Thus pinpointing the effects of pollution can be very difficult. And while considerable progress has been made in understanding how individuals value improved health and other welfare enhancements, there is still great uncertainty attached to many values. Benefit estimation is therefore far from straightforward.

The estimation of costs is generally assumed to be simpler, since it appears on the surface to involve a mere toting up of the expenditures made by regulatees for pollution control equipment, less-polluting fuels, and the like. In actuality, estimating the social costs of regulation is much more complex.

In economics the true social cost of any policy, regulatory or otherwise, is measured by the total amount of compensation required to leave all affected consumers as well off after a policy goes into effect as before. Thus if a regulation makes some products more expensive, the total compensation would include the amount necessary to enable consumers to buy the products at their higher prices. If a regulation results in involuntary unemployment, the compensation required would include the lost income as well. It would also include losses to consumers if and when products disappear from the market because a regulation makes them too expensive to produce profitably. Only if costs are estimated in this fashion will the resulting numbers be appropriate to compare with benefit estimates based on willingness to pay.

In principle the direct pollution control expenditures by regulatees—that is, the amounts they spend for pollution abatement equipment, cleaner fuels, and the like—could reasonably approximate the correct (compensation-based) measure of social cost.[24] In practice two problems complicate the issue, however.

First, the conditions under which expenditures can fairly be said to approximate true social costs are not very realistic. They require that the markets for all final goods and services, as well as factors of production, be perfectly competitive. In addition, there can be no substitution in either production or consumption if expenditures are to accurately measure true social costs. This is even less realistic.

Second, even if these conditions were met, so that expenditures did proxy social costs, predicting expenditures in advance of a proposed regulation is not at all straightforward. For one thing, even prospective regulatees themselves are often uncertain about how they will go about meeting a standard when specific control equipment is not required. In practice, meeting a standard may be more or less expensive than they estimate initially. Beyond that, some costs get passed along to other industries that might appear at first blush to be unaffected by a regulation. For instance, air pollution control requirements for the electric utility industry will increase costs in industries like banking and communications, even though the latter produce very little air pollution and would not be required to make capital investments in pollution control. Thus one has to look beyond initial impacts. Finally, even when prospective regulatees have a good idea of what required control expenditures will be, they have an incentive to overstate expenditures in the hope that the chances of regulation may be lessened.[25]

For all these reasons, then, expenditure estimates are generally poor proxies for the true costs of air pollution control. Until recently, however, studies claiming to show the costs of pollution control actually presented an amalgam of expenditure estimates coupled with true costs in the economic sense. (Several of these studies, as well as a recent and much more careful attempt, are reviewed in the section on Expenditure and Cost Estimates below.)

Benefit Estimates

Comprehensive studies. By far the most ambitious attempt at estimating air pollution control benefits is that of A. Myrick Freeman. First in a report to the Council on Environmental Quality and subsequently in a book based on that report,[26] he surveyed virtually all previous benefit studies and attempted to synthesize them into an overall estimate. Because it draws on such a large body of work, Freeman's study is discussed in some detail here.

Even Freeman's serious effort, however, is not fully comprehensive in the sense discussed in the preceding section on fundamentals. Freeman made no attempt to determine the location and magnitude of the emission reductions resulting from each of the regulations written under the Clean Air Act. Nor did he attempt independently to link these emission reductions to improvements in ambient air quality. Rather, Freeman elected to estimate the national benefits associated with the *observed* changes in air quality between 1970 and 1978. These amounted to about a 20 percent reduction in TSP and SO_2 concentrations, a 30 percent reduction in CO, virtually no change in O_3, and an insufficient basis for judging changes in NO_2. To the extent these improvements were directly attributable to the Clean Air Act, of course, this approach was tantamount to estimating the benefits of the act itself. For reasons mentioned above, however, such an attribution is risky.[27]

As the next step, these realized improvements in air quality had to be translated into increases in individuals' well-being. Freeman considered three categories of benefits: those to human health (from reduced morbidity and mortality); those resulting from reduced damage to vegetation and exposed materials, and from reduced cleaning costs; and those amenity benefits primarily taking the form of improved visibility. In order to estimate the benefits in each of these categories, Freeman drew exclusively on previous studies designed to identify such links; that is, he did not himself attempt to estimate the relationship(s) between, say, ambient particulate concentrations and respiratory disease or between ozone and agricultural output.

One problem with this approach is that each such relationship (as that between SO_2 and premature mortality, for instance) has been explored many times and in a variety of different ways, and different studies have come to very different conclusions. Some find one or more air pollutants to be serious health hazards, while others find little or no evidence linking these same pollutants to adverse health. Even when studies are in agreement that a particular pollutant may be harmful to humans or ecosystems, they may differ as to the concentration at which damage begins (or whether such a threshold exists at all). Thus Freeman presented his estimates in the form of a range where the lower bound generally reflects studies which found little or no pollution damage to health, vegetation, and the like, while the upper bound reflects those studies finding the most significant effects. He also made a mid-range or "most reasonable point estimate," reflecting his "best guess" and based on the quality of the various studies surveyed.

Freeman followed a similar strategy in valuing the benefits resulting from air quality improvements. That is, he surveyed a wide range of studies which derived values to be attached to reduced illness, crop loss, and materials damage, and even to the reductions in premature mortality that might result from reduced air pollution. Some of these values are rather straightforward. If improved air quality means ten more bushels of soybeans can be grown in the United States, the effect is easy to value since soybeans trade at well-defined market prices. Attaching a value to lives "saved" (actually just prolonged) through air pollution control is much more complex; current estimates of the value of what economists call a statistical life—meaning one individual whose identity is unknown—range from about $1 million to as much as $8 million (in 1984 dollars).[28] Freeman used $1 million in his estimates, although he invited readers to substitute higher or lower values if they wished.[29]

Finally, Freeman attempted to estimate benefits separately for stationary and mobile source controls. This he did by ascribing to stationary source controls the benefits from reduced TSP and SO_2 concentrations, while any benefits resulting from reduced O_3, CO, or NO_2 were attributed to mobile source controls. This has the advantage of allowing a comparison of benefit-cost ratios for each program individually, but is also misleading since three-fifths of all hydrocarbon and NO_2 emissions (precursors of ozone) and nearly a third of all CO emissions come from stationary sources.[30] Thus a fair share of the benefits Freeman ascribes to mobile source pollution control may be due to controls on factories, power plants, and other stationary sources. Table 3–5 presents Freeman's findings.

There are a number of important observations to be drawn from Freeman's estimates, the most important of which concerns the uncertainty about the "true" benefits. According to Freeman, total air pollution control benefits in 1978 could have ranged from as little as $8.8 billion to as much as $91.6 billion—a difference greater than an order of magnitude. The range for total health benefits—$5.0 to $64.9 billion—differs by a factor of thirteen. These ranges reflect the great diversity of opinion, based on different research studies, about the effects of air pollution on health and other benefit components. In fact, the lower bound of Freeman's health benefit estimates probably should be even lower, since other credible studies conclude that the criteria air pollutants have *no* adverse effects on health *at present levels in most places* (there is no doubt that at elevated concentrations all the criteria air pollutants are harmful).

Table 3–5. Benefits in 1978 from Air Quality Improvements Between 1970 and 1978 (billions of 1984 dollars)

	Stationary source (TSP, SO_2)		Mobile source (O_3, CO, NO_2)		Total	
Benefit category	Range	Most reasonable point estimate	Range	Most reasonable point estimate	Range	Most reasonable point estimate
Health	$ 5.0–64.3	$27.2	$ 0.0–0.6	$ 0.0	$ 5.0–64.9	$27.2
Soiling and cleaning	1.6–9.6	4.8			1.6–9.6	4.8
Vegetation	0.0	0.0	0.2–0.6	0.5	0.2–0.6	0.5
Materials	0.6–1.8	1.1	0.0–0.5	0.0	0.6–2.3	1.1
Property values (aesthetics)	1.4–11.0	3.7	0.0–3.2	0.0	1.4–14.2	3.7
Total	8.6–86.7	36.8	0.2–4.9	0.5	8.8–91.6	37.3

Note: 1978 dollars have been converted to 1984 dollars using the consumer price index.
Source: Adapted from A. Myrick Freeman III, *Air and Water Pollution Control: A Benefit-Cost Assessment* (New York, Wiley, 1982) p. 128.

Another interesting aspect of Freeman's estimates is the overwhelming importance of health benefits in the total. According to his most reasonable point estimates, nearly three-fourths of the total air pollution control benefits are due to reduced sickness and reduced premature mortality. The next largest component, reduced soiling and cleaning costs, accounts for a little more than one-eighth of the total. These results would appear to support the emphasis in the Clean Air Act on standards to protect human health.

A third quite striking finding in Freeman's analysis is the allocation of total benefits between the stationary and mobile source control programs. Again using the most reasonable point estimates, 99 percent of the total benefits Freeman finds are due to the control of particulates and sulfur dioxide—both overwhelmingly stationary-source pollutants. According to Freeman, mobile source benefits are not likely to exceed $5 billion, a figure well below the lower bound estimate for stationary source control benefits, $8.6 billion. On the basis of this study it would appear that the United States has gotten relatively little from its investment in mobile source pollution control.

The latter observation must be strongly qualified. Recall that the basis for Freeman's estimates was actual improvements in air quality between 1970 and 1978. Although mobile source pollution controls may not have effected substantial air quality improvements for either ozone or nitrogen dioxide (thus providing few benefits for Freeman to calculate), they may have prevented further deterioration in air quality. This was quite probably the case, since both the number of vehicles and the number of vehicle miles traveled increased by 60 percent between 1967 and 1979.[31] Keeping air quality constant during such an increase in vehicle use was an accomplishment. Thus if Freeman had based his calculations on the difference in air quality in a world with controls and one without controls, his estimates of mobile source control benefits would have been higher. In addition, the benefits from control of fine particulates from diesel-fueled cars, trucks, and buses were not included in Freeman's estimates because these controls were introduced after 1978. Finally—and of some importance—Freeman's estimates included no benefits from reductions in ambient concentrations of lead. This is probably a significant omission, since the Clean Air Act has resulted in substantially reduced lead concentrations in many places and since lead may have serious adverse effects on health, particularly in children.

What other qualifications or observations can we offer about the findings of this rare effort to identify in a comprehensive way the benefits of air pollution control? First, as Freeman is quite clear in pointing out, his estimates include only benefits from controlling the common air pollutants. If other, more exotic pollutants like benzene or beryllium are controlled by the same equipment that removes TSP or SO_2, say, and if significant human exposures are thus prevented, then true health benefits may be larger than Freeman estimates.[32]

Second, since the time Freeman's analysis was done several new studies have been completed which suggest possible substantial benefits from the prevention of air pollution damage to agricultural crops and to exposed materials.[33] Had these studies been available when Freeman synthesized existing work, he might have identified a higher upper bound in those benefit categories and perhaps a higher mid-range estimate. This might have been so particularly for damages caused by TSP or SO_2, since—it is emphasized once again—his estimates were based solely on realized improvements between 1970 and 1978, which were greatest for those pollutants.

Third, Freeman's estimates of aesthetic benefits may be open to question (indeed, he expresses some reservations about the property value approach to estimating such benefits). His mid-range estimate of $3.7 billion in 1978 seems low in view of several recent but preliminary studies that estimate visibility benefits not by the property value approach, but rather by asking people directly how much they would be willing to pay for improved visibility.[34] Had these studies been available, Freeman's estimates of aesthetic benefits might have been higher.

The preceding observations all point to ways in which Freeman's estimates may be on the low side. Are there respects in which they may err in the opposite direction? There are several. To begin with, his estimates reflect Clean Air Act benefits only to the extent that the air quality improvements between 1970 and 1978 were due to this statute. Since some of those improvements seem likely to have resulted from purely economic forces or independent state regulation, total benefits would have to be adjusted downward. Next, a substantial portion of the health benefits Freeman identifies were due to reduced mortality. As was observed above, each life saved (or death postponed) was valued at $1 million, a figure derived from studies of risk valuation by people in the prime of their working lives. However, it seems likely that if air pollution affects the annual mortality rate, it does so primarily at the upper end of the age distribution—that is, it extends the lives of the elderly by several years rather than reducing the death rate in all age groups.[35] If, as seems possible, society values less highly the prolongation of the life of an elderly (and perhaps infirm) person than it does the life of a healthy middle-aged one, Freeman's estimate would be biased upward.

Recently a contractor to the Environmental Protection Agency attempted to update Freeman's findings, filling in where possible some of the gaps identified above.[36] Among other things, the update built upon recent information developed by the EPA on the benefits of controlling ambient lead concentrations; recent studies of the health damages and agricultural damages associated with exposures to ambient ozone; and estimates of population-weighted changes in ambient concentrations of TSP and SO_2 (which take into account the fact that the largest reductions in concentrations of these two pollutants have occurred in heavily populated parts of the country). However, for the most part the contractor's study adhered to Freeman's approach of basing benefit estimates on realized improvements in air quality since 1970.

The contractor's study, which must be regarded as preliminary, concluded that by 1981 aggregate benefits from air pollution control in the United States were in the range of $22.7 to $62.1 billion. No mid-range or best estimate was offered. Thus the update arrived at a much higher lower bound than Freeman did ($22.7 billion versus $8.8 billion), but also posited a lower value for the likely upper bound estimate ($62.1 billion versus Freeman's $91.6 billion) since different studies were used to establish that bound.

Other studies. Although Freeman's study and the effort to update it are the only recent comprehensive estimates of the benefits of air quality improvements, there exist other, more narrowly focused studies that deserve mention.

A study by Mathtech, Inc., concentrated exclusively on airborne particulate matter and the benefits that would result from hypothetical changes in the current TSP standard.[37] Thus this study was not designed to estimate benefits associated with existing regulations under the Clean Air Act, but rather to assess possible future benefits resulting from more stringent regulation. (Virtually all other benefit studies are of this type; that is, they examine the consequences of changes in the current status quo as opposed to the benefits associated with existing regulations.)

The Mathtech study attempted to answer the following question: What would be the present value (in 1982) of the benefits that would arise during the period 1989 to 1995 from a variety of hypothetical standards for particulate matter that would improve air quality during that period? Among other things, the study required projections of the numbers of individuals and the kinds of materials likely to be exposed during that period; the air quality improvements that would result from each hypothetical standard; a translation of these changes into improvements in health, reduced soiling and materials damage, and other improvements in well-being; and the dollar valuation of these changes. To accomplish the latter two tasks an approach not unlike Freeman's was employed; that is, a wide range of studies was surveyed, each of which either estimated or assigned dollar values to the effects of air pollution on health, materials, and the like. Then methods were developed for averaging the results and obtaining "consensus" estimates of benefits in various categories. Also like Freeman's, this study reflected many of the uncertainties inherent in such an approach by presenting its findings in the form of a range, with upper and lower bounds and a mid-range point estimate.

Table 3–6 presents estimates of the benefits associated with one of the ten hypothetical particulate matter standards considered in the study discussed above.[38] Columns A–F indicate either differing views of the kinds of benefits to be included in the total estimate or the findings of specific studies upon which the estimates in each benefit category are based. Thus under scenario A—the most conservative of those considered—no benefits are recognized from possible reductions in acute morbidity or household cleaning costs, and estimates of benefits from reduced mortality are based on studies that find fairly small effects. In scenarios B and C, benefits are added from reduced morbidity as well as reduced soiling and cleaning. Beginning with scenario D, mortality benefits rise because different studies are used as the basis for calculations. Thus, as one moves from scenario A through F, benefits grow both because more categories are included and because different studies are used as the bases for the estimates. Scenario F, then, is the most liberal and inclusive of those examined.[39]

Another interesting study, conducted by the EPA, dealt with the benefits of reducing the maximum allowable lead content of gasoline, which the EPA has regulated since 1973 under the Clean Air Act.[40] At the beginning of 1985 the allowable limit stood at 1.1 grams per gallon (g/gal) of leaded gasoline. In 1985, however, the agency issued a regulation reducing the limit to 0.5 g/gal, effective in July 1985, and to 0.1 g/gal effective in January 1986. The purposes of the rulemaking were to reduce ambient airborne concentrations of lead, and to reduce the price difference between leaded and unleaded gasoline and thus to discourage the use of leaded gas in catalyst-equipped cars (such misfueling can poison the catalytic converter used to reduce emissions of hydrocarbons, carbon monoxide, and nitrogen oxides). In support of the change in the lead limit, the EPA prepared an unusually detailed analysis of the costs and benefits of the change.

According to that analysis, four types of benefits would result from the lead phase-down, as it was called: (1) reduced incidence, among children, of adverse health and cognitive effects associated with lead; (2) reduced incidence, in adult males, of blood-pressure-related effects due to lead; (3) fewer health and other damages resulting from emissions of hydrocar-

Table 3–6. Benefits from the Imposition of a Tighter Particulate Standard (billions of 1984 dollars)

	Benefit scenario[a]					
Benefit category	A	B	C	D	E	F
Mortality	$1.41	$1.41	$ 1.41	$16.03	$16.03	$17.44
Acute morbidity	0.00	1.66	13.42	13.42	13.42	15.08
Chronic morbidity	0.15	0.15	0.15	0.15	14.36	14.36
Housing cleaning and materials damage	0.00	0.00	0.92	0.92	3.96	17.45
Manufacturing cleaning and materials damage	0.00	0.00	0.00	0.00	1.64	1.64
Total	1.56	3.23	15.91	30.54	49.42	65.97

Note: The table gives present discounted value (at 10 percent) in 1982 of benefits enjoyed 1989–1995. Column totals may not add up due to rounding.

Source: Adapted from Mathtech, Inc., "Benefit and Net Benefit Analysis of Alternative National Ambient Air Quality Standards for Particulate Matter," report for the Economic Analysis Branch, Office of Air Quality Planning and Standards, Environmental Protection Agency (March 1983) p. 1-52.

[a] For definitions of scenarios, see Mathtech, Inc., "Benefit and Net Benefit Analysis," p. 1-44.

bons, carbon monoxide, and nitrogen oxides from misfueled cars; and (4) reduced automotive maintenance expenditures (according to the EPA, cars would require fewer tune-ups and oil changes and their exhaust systems would last longer). To value the health benefits to children and adults, the EPA made assumptions about the medical cost of treating children having elevated levels of lead in the blood; the compensatory education required for children whose school performance is affected by lead-related impairments; the medical expenditures required of hypertensives and of heart attack and stroke victims (benefits of reduced mortality were valued at $1 million per statistical life saved); and health costs arising from exposure to ozone, carbon monoxide, and nitrogen oxides.

Table 3–7 presents the EPA's findings in summary form. Note that the net benefits (total benefits minus total costs) associated with reduced lead in gasoline are positive in all years considered and are quite substantial as well. They are largest in 1986 (the year in which the maximum permitted lead level was reduced to 0.1 g/gal) and decline slowly with time as fewer and fewer cars using leaded gasoline remain in use. Even if the benefits associated with elevated blood pressure in adult males are excluded,[41] net benefits are still positive and substantial. Thus the EPA's benefit-cost analysis strongly supports its rulemaking on lead in gasoline. This is worth noting since some critics seem to believe that such analyses are only used to oppose federal regulations.

The benefit studies discussed here were selected for their careful execution and because they were comprehensive or dealt with important pollutants or industries. Nevertheless, they are but a few examples of the large and growing number of studies estimating air pollution control benefits. It is important to realize that in spite of the difficulties that arise, benefit estimation is a useful input in policymaking when used carefully.

Expenditure and Cost Estimates

Recall the distinction drawn above between direct expenditures necessitated by regulation on the one hand and the true social costs of regulation on the other. Previous studies of the costs of air quality regulation have actually combined expenditure data with some true costs as economists might measure them. Although we will refer to these as cost estimates, we will use quotation marks on the word "cost" to indicate that it really represents a hybrid.

From 1972 through 1980, the White House Council on Environmental Quality (CEQ) was the major source of information on environmental spending pursuant to regulation. The CEQ's Annual Report to the President contained estimates for spending under all federal environmen-

Table 3–7. Year-by-Year Costs and Monetized Benefits of Lead Regulation, Assuming Partial Misfueling, 1985–1992
(millions of 1984 dollars)

	1985	1986	1987	1988	1989	1990	1991	1992
Benefit category	$	$	$	$	$	$	$	$
Children's health effects	232	623	568	521	470	430	383	372
Adult blood pressure	1,790	6,124	5,893	5,657	5,387	5,157	4,862	4,872
Conventional pollutants	0	231	231	233	235	239	248	258
Maintenance	106	949	892	849	818	797	783	779
Fuel economy	36	194	177	117	139	144	179	170
Total monetized benefits	2,164	8,121	7,761	7,377	7,049	6,767	6,455	6,449
Total refining costs	100	631	579	552	523	489	461	458
Net benefits	2,065	7,490	7,181	6,825	6,526	6,278	5,994	5,991
Net benefits excluding blood pressure	274	1,366	1,288	1,168	1,139	1,121	1,132	1,119

Source: Office of Policy Analysis, Environmental Protection Agency, *Costs and Benefits of Reducing Lead in Gasoline*, Report no. EPA-230-05-85-006 (February 1985) p. E-12.

tal statutes, including the Clean Air Act. Unfortunately, this practice was discontinued in 1981 and no such estimates have appeared from the council since. Table 3–8 below presents the CEQ's estimate of air pollution expenditures in 1979 (the last year for which an estimate was made).

According to the council, air pollution expenditures under the Clean Air Act amounted to more than $30 billion in 1979. This was divided evenly between capital "costs"—interest charges and depreciation on pollution

Table 3–8. CEQ Estimate of Air Pollution Control Expenditures for 1979 and for 1979–1988
(billions of 1984 dollars)

Sector	Total annual expenditures, 1979	Projected cumulative expenditures, 1979–1988
Governments	$ 2.15	$ 27.89
Industrial	6.26	84.08
Electric utilities	12.02	150.15
Mobile sources	11.59	165.59
Total	31.19	427.71

Note: 1979 dollars have been converted to 1984 dollars using the consumer price index.

Source: Adapted from Council on Environmental Quality, *Environmental Quality: 1980* (Washington, D.C., Government Printing Office, 1980) p. 394.

control equipment—and the cost of operating and maintaining that equipment. The electric utility industry accounted for nearly 40 percent of the total, while controls on automobiles and other mobile sources accounted for another 37 percent. All other industries taken together accounted for less than a fifth of total air pollution expenditures as defined by the CEQ.

In 1984 a somewhat different estimate was released by the Environmental Protection Agency, which is required periodically to report Clean Air Act and Clean Water Act compliance "costs" to Congress. Table 3–9 presents the EPA's estimates in a form comparable to those of the council in table 3–8.

A side-by-side comparison of tables 3–8 and 3–9 shows how widely such estimates can differ (although note that the differences are typically narrower than the ranges in the benefit estimates). The EPA estimates of both one-year and decadal air pollution control expenditures are significantly lower than those of the CEQ. This is so even though the EPA estimates pertain to later periods (1981 versus 1979 and 1981–1990 versus 1979–1988), which should, if anything, make them higher because of new regulatory requirements imposed between 1979 and 1981. The most significant differences between the two sets of estimates involve the electric utility industry and mobile sources, two of the most heavily regulated sources of conventional air pollutants.

Table 3–9. EPA Estimate of Expenditures Necessitated by Air Pollution Control Regulations, for 1981 and for 1981–1990 (billions of 1984 dollars)

Sector	Annual cost, 1981	Cumulative costs, 1981–1990
Governments[a]	$ 0.72	$ 6.29
Industrial[b]	7.21	84.67
Electric utilities	8.49	109.08
Mobile sources	6.90	91.83
Total	23.32	291.87

Source: Adapted from Office of Policy Analysis, Environmental Protection Agency, "The Cost of Clean Air and Water Report to Congress, 1984," Report no. EPA-230-05-84-008 (May 1984) pp. 11–12.

[a] Represents the sum of Government Expenditures and Municipal Waste Incineration entries in the EPA report.

[b] Represents all entries other than electric utilities, mobile sources, and government and municipal waste incineration.

Which set of numbers is most accurate? It is difficult to say, in part because the methodologies behind each estimate are not spelled out very clearly. The EPA report does contain some discussion of its approach to "cost" estimation, but it often raises more questions than it answers. The CEQ estimates have always been more ad hoc in nature, relying in part on EPA studies, in part on studies done outside the government, and in part on original estimates for certain sectors or categories of sources.[42]

Nevertheless, it may be possible to draw some conclusions about "best" estimates. For instance, the EPA's estimate of air pollution "costs" for the electric utility industry was based in part on a comprehensive report to the agency on environmental regulation in that industry that was unavailable at the time the CEQ estimate was made.[43] Thus the EPA's estimate is probably the more accurate for the electric utility industry circa 1980. On the other hand, two recent and independent studies give reason to believe that "costs" associated with the mobile source emission controls in the Clean Air Act are much greater than estimated by the EPA, and perhaps even greater than estimated by the CEQ. One recent study by the Brookings Institution of federal efforts to control mobile source pollution reviewed all previous "cost" estimates.[44] On the basis of that review, it concluded that the mobile source standards in place in 1981 would entail a lifetime "cost" of $1,650 per 1981 vehicle and a comparable sum for light-duty trucks if the latter were required to meet a 75 percent reduction in emissions of nitrogen oxides by 1985. Lifetime "costs" for heavy-duty trucks would be even greater according to the study. In the aggregate, it was estimated that in a steady state (when all vehicles are meeting the 1981 standards), the total annual "cost" of the mobile source controls in place in 1981 would be $18 billion.[45]

More recently, the Bureau of Economic Analysis (BEA) of the Department of Commerce updated its estimates of spending for emissions control from mobile sources. Like the Brookings study, the BEA estimates include not only hardware "costs" (for catalytic converters, for example) but also fuel economy penalties and higher operating and maintenance costs. According to the bureau, in 1981 the nation was spending $15.2 billion to abate air pollutant emissions from all vehicles, $8.6 billion of which was for equipment with another $6.6 billion going to higher operating "costs."[46] By 1984, incidentally, total "costs" had risen to $17.8 billion, according to the BEA. On the basis of the latter two studies, it would appear that mobile source air pollution "costs" in 1981 were closer to $15 billion (perhaps more) than to the EPA's estimated $7 billion.

In 1986, Hazilla and Kopp made the first real attempt to estimate the true social costs associated with existing environmental regulations.[47] To do so they began with EPA estimates of initial expenditures by regulatees, not unlike those presented in table 3–9. They then made two important modifications. First, using a general equilibrium model of the economy they developed, they traced through to their ultimate incidence these initial pollution control expenditures incurred in the manufacturing and other sectors of the economy as a result of regulations. Second, using data on actual consumer expenditure patterns and making an assumption about the nature of consumer preferences, Hazilla and Kopp evaluated all the resulting price and income changes predicted by the model in terms of compensation required to keep consumers as well off after the regulations were imposed as they would have been without them. Thus their estimates are of pollution costs in the true economic sense, and therefore are comparable to benefit estimates like those discussed in the preceding section.

One problem arises in using the findings from this study. Hazilla and Kopp did not analyze air and water pollution control regulations separately, but rather took as their initial point of departure EPA estimates of initial pollution expenditures for air and water pollution control combined, on an industry-by-industry basis. Only if we make a simplifying assumption—that the true social costs of air and water pollution control will be in proportion to the initial pollution control expenditures—can we isolate the air pollution component. Under that assumption, in table 3–10 the EPA estimates of air pollution control "costs" in three selected years are compared with the Hazilla and Kopp estimates of true costs.

As table 3–10 indicates, estimating the costs associated with air pollution control in the conceptually correct manner makes a huge difference. Compared to the more crude EPA estimates, the true social costs of complying with federal air pollution control regulations were significantly less through the early 1980s. By 1985, however, social costs (as estimated by Hazilla and Kopp) exceeded the EPA's estimate by nearly a third ($33.0 billion versus $24.6 billion).

What accounts for these sharp differences? One possible explanation follows. In the early years of federal air quality regulation, new rules were issued sequentially on an industry-by-industry basis. This meant that as the first industries were regulated, some substitution in production to unregulated inputs and intermediate products was possible. Similarly, consumers could relatively easily substitute among final products, at least some of which had yet to go up in price because of regulation. Over time, however, air quality regulation came to be more comprehensive, thus making such substitution more difficult. The gap between true social costs and more crude expenditure-based estimates began to close.

Equally important, over time significant amounts of investment capital began to be diverted from productivity-enhancing uses to air quality protection.[48] This had the effect of shifting the economy onto a lower growth path, and by 1985 the reduction in the rate of growth had become significant enough that true social costs—as measured by compensation

Table 3–10. Comparison of EPA Air Pollution Control "Cost" Estimates with Hazilla-Kopp Estimates for True Social Costs (billions of 1984 dollars)

Year	EPA	Hazilla-Kopp
1975	12.7	4.5
1981	20.5	13.7
1985	24.6	33.0

Sources: Author's estimates based on Michael Hazilla and Raymond J. Kopp, "The Social Cost of Environmental Quality Regulations: A General Equilibrium Analysis," Discussion Paper QE86-02, Resources for the Future (August 1986); and Office of Policy Analysis, Environmental Protection Agency, "The Cost of Clean Air and Water Report to Congress, 1984," Report no. EPA-230-05-84-008 (May 1984) pp. 1-8.

required—had come to exceed the EPA's estimates by a significant margin.

As Hazilla and Kopp are careful to point out, their estimates must be regarded as quite preliminary. Among other things, the general equilibrium model they use is crude in several respects; they are forced to rely for a starting point on EPA expenditure estimates of sometimes questionable quality; and the calculation of compensation required is based on outdated data on patterns of household expenditures and on an assumed form of consumer preferences. Nevertheless, theirs is the first attempt to estimate air and water pollution control costs correctly, and theirs are the only cost estimates deserving of comparison with the benefit estimates in the preceding section.

Benefit-Cost Comparisons

It is tempting to leap from the preceding two sections to comparisons of benefits and costs. After all, it is important to know whether the gains from air pollution control efforts are worth what society must give up to have them. Indeed, that is one reason—perhaps the major reason—why benefits and costs are estimated.

The best we can do here is to make a rough comparison. Hazilla and Kopp estimate that the total social cost of air pollution control regulations was $13.7 billion in 1981. According to Freeman, by 1978 the best estimate of the benefits associated with improved U.S. air quality since 1970 was $37.3 billion. Freeman's estimate would have been even higher for the year 1981 because the U.S. population was greater, and the air quality improvements even larger as measured from the 1970 base year. Thus one is tempted to conclude that as of 1981, at least, the benefits of air pollution control were considerably in excess of the costs.

That conclusion is probably correct for 1981, but it should be qualified in several respects. First, it should be noted that according to Hazilla and Kopp, air pollution control costs had climbed sharply from 1981 to 1985, much faster than benefits could have been expected to increase. Thus what was perhaps true in 1981 may not be true today. Second, recall that Freeman attributed all of the improvements in air quality between 1970 and 1978 to the Clean Air Act; for reasons discussed above, this probably exerts an upward bias, although in other respects Freeman's estimates are probably conservative. Third, note that although the total benefits of air pollution control may exceed the total costs, we could still be overregulating if, at the margin, we could save more in control costs by relaxing regulations slightly than we would sacrifice in additional health and other damages. Since the marginal cost of air pollution control increases sharply as the level of control increases, while marginal benefits are likely to remain constant or even decline, this is quite likely to be the case. Thus while the Clean Air Act is probably a very desirable policy on a take-it-or-leave-it basis, it is worth looking carefully at possible adjustments in the levels of control it calls for.

Cost-Effectiveness Analysis and Air Pollution Control

Because of difficulties in identifying and valuing the favorable effects of air pollution control (particularly reduced mortality and morbidity), benefit-cost comparisons will always be controversial. Indeed, such comparisons generally do not play much of a role in establishing the goals of air pollution control or other environmental policies. These goals, as embodied in legislation, usually are the result of political compromises between environmental and industry groups, state and local officials, the executive branch and various members of Congress, and other interested and politically active parties.

That this is so does not vitiate the potential contribution of analytical methods to air pollution policy. These methods can play a large role in helping society meet predetermined environmental or other goals as inexpensively as possible—an approach referred to as cost-effectiveness analysis. Such analyses of air pollution policy have come to much less ambiguous conclusions than those attempting to compare costs and benefits. Specifically, a very large body of research has demonstrated clearly that existing limits on pollutant emissions, and perhaps current air quality goals, could be met at a fraction of what is now being spent.

The studies reaching this conclusion generally employ a common methodology. Typically, they search for possible variations in the cost of controlling the last ton(s) of air pollutants removed at regulated sources. Where variations exist, cost savings are possible. For instance, if a coal-fired power plant is spending $2,000 per ton to remove the last several tons of SO_2, but a nearby smelter could remove an additional ton for $1,000, total control costs could be reduced by $1,000 by allowing the power plant to emit an additional ton (thereby saving $2,000) while requiring the smelter to remove an additional ton (for $1,000). Emissions remain the same, control costs fall. If this process were carried to its logical conclusion, total control costs would be minimized when all regulated sources faced the same marginal cost of control on the last unit of pollution removed. The same logic suggests that controls on multiple sources (or stacks) at one plant could be manipulated to minimize control costs for that plant.

In part to determine the savings possible through more cost-effective air quality regulation, Tietenberg recently surveyed the most important of these studies.[49] His findings are summarized in table 3–11; the column labeled "CAC [command and control] benchmark" indicates the regulatory approach against which cost savings were measured. Thus in the study by Roach and coauthors, actual SIP (state implementation plan) regulations on SO_2 emissions in four southwestern states were compared to a least-cost solution. Similarly, Maryland SIP regulations formed the basis for McGartland's study of TSP control in Baltimore, although he focused on ambient air quality goals rather than emissions goals. The ratios in the right-hand column show how much more expensive the traditional (command-and-control) regulatory approach is when compared to the most cost-effective set of controls: the range is from a low of 7 percent more expensive in a study of SO_2 control in the Los Angeles basin up to 2,100 percent more expensive in a study of hypothetical NO_2 regulations in Chicago. It appears from Tietenberg's survey that traditional air quality regulation may be, on average, about three or four times more expensive than the most cost-effective approach—if these studies are representative of real-world experience.

Tietenberg cites several reasons why actual savings may be less than suggested by the studies he reviewed. The most important is that pollution control equipment is already in place at many facilities because of existing regulations, and could not be costlessly dismantled and reinstalled at those sources that would be controlled under a cost-minimizing approach.[50] A problem could also arise under a least-cost approach if one source in an area were allowed to increase its pollution substantially in exchange for reductions from a number of other sources spread throughout that area. In such a case, even though pollution might remain the same in the aggregate, it would be much more heavily concentrated than before at one location. For this reason, many cost-minimizing schemes include provisions to prevent hot-spot problems of this kind. These provisions would reduce the cost savings possible from a reallocation of the control burden.

These and other qualifications notwithstanding, the studies that Tietenberg reviews (and others as well) give some idea of the kind of savings that would be possible through more selective regulation. Suppose, for example, that command-and-control regulation were 50 percent more expensive than more cost-effective rules (rather than 200 to 300 percent, as suggested in table 3–11). If the social cost estimates from the preceding section of this chapter are at all accurate, it would have been possible to save $7 billion in 1981 alone through wiser air quality regulation, with

Table 3–11. Empirical Studies of Air Pollution Control

Study and year	Pollutants covered	Geographic area	CAC benchmark	Assumed pollutant type	Ratio of CAC cost to least cost
Atkinson and Lewis (1974)	Particulates	St. Louis metro. area	SIP regulations	Nonuniformly mixed assimilative	6.00[a]
Roach and coauthors (1981)	Sulfur dioxide	Four Corners in Utah, Colorado, Arizona, and New Mexico	SIP regulations	Nonuniformly mixed assimilative	4.25
Hahn and Noll (1982)	Sulfates	Los Angeles	California emission standards	Nonuniformly mixed assimilative	1.07
Krupnick (1983)	Nitrogen dioxide	Baltimore	Proposed RACT regulations	Nonuniformly mixed assimilative	5.96[b]
Seskin, Anderson, and Reid (1983)	Nitrogen dioxide	Chicago	Proposed RACT regulations	Nonuniformly mixed assimilative	14.4[b]
McGartland (1984)	Particulates	Baltimore	SIP regulations	Nonuniformly mixed assimilative	4.18
Spofford (1984)	Sulfur dioxide	Lower Delaware Valley	Uniform percentage reduction	Nonuniformly mixed assimilative	1.78
	Particulates	Lower Delaware Valley	Uniform percentage reduction	Nonuniformly mixed assimilative	22.0
Harrison (1983)	Airport noise	United States	Mandatory retrofit	Uniformly mixed assimilative	1.72[c]
Maloney and Yandle (1984)	Hydrocarbons	All domestic du Pont plants	Uniform percentage reduction	Uniformly mixed assimilative	4.15[d]

Palmer, Mooz, Quinn, and Wolf (1980)	Chlorofluorocarbon emissions from nonaerosol applications	United States	Proposed emission standards	Uniformly mixed accumulative	1.96

Definitions: CAC = Command and control, the traditional regulatory approach
SIP = State implementation plan
RACT = Reasonably available control technologies, a set of standards imposed on existing sources in nonattainment areas

Sources: T. H. Tietenberg, *Emissions Trading: An Exercise in Reforming Pollution Policy* (Washington, D.C., Resources for the Future, 1985), citing the following sources: Scott E. Atkinson and Donald H. Lewis, "A Cost-Effectiveness Analysis of Alternative Air Quality Control Strategies," *Journal of Environmental Economics and Management* vol. 1, no. 3 (November 1974) p. 247; Fred Roach, Charles Kolstad, Allen V. Kneese, Richard Tobin, and Michael Williams, "Alternative Air Quality Policy Options in the Four Corners Region," *Southwest Review* vol. 1, no. 2 (Summer 1981) table 3, pp. 44–45; Robert W. Hahn and Roger G. Noll, "Designing a Market for Tradeable Emission Permits," in Wesley A. Magat, ed., *Reform of Environmental Regulation* (Cambridge, Mass., Ballinger, 1982) tables 7–5 and 7–6, pp. 132–133; Alan J. Krupnick, "Costs of Alternative Policies for the Control of NO_2 in the Baltimore Region" (Resources for the Future working paper, 1983) table 4, p. 22; Eugene P. Seskin, Robert J. Anderson, Jr., and Robert O. Reid, "An Empirical Analysis of Economic Strategies for Controlling Air Pollution," *Journal of Environmental Economics and Management* vol. 10, no. 2 (June 1983) tables 1 and 2, pp. 117 and 120; Albert M. McGartland, "Marketable Permit Systems for Air Pollution Control: An Empirical Study" (Ph.D. dissertation, University of Maryland, 1984) table 4.2, p. 67a; Walter O. Spofford, Jr., "Efficiency Properties of Alternative Source Control Policies for Meeting Ambient Air Quality Standards: An Empirical Application to the Lower Delaware Valley" (Resources for the Future Discussion Paper D-118, February 1984) table 13, p. 77; David Harrison, Jr., "Case Study 1: The Regulation of Aircraft Noise," in Thomas C. Schelling, ed., *Incentives for Environmental Protection* (Cambridge, Mass., MIT Press, 1983) tables 3.6 and 3.16, pp. 81 and 96; Michael T. Maloney and Bruce Yandle, "Estimation of the Cost of Air Pollution Control Regulation," *Journal of Environmental Economics and Management* vol. 11, no. 3 (September 1984) pp. 244–263; Adele R. Palmer, William E. Mooz, Timothy H. Quinn, and Kathleen A. Wolf, *Economic Implications of Regulating Chlorofluorocarbon Emissions from Nonaerosol Applications*, Report #R-2524-EPA, prepared for the U.S. Environmental Protection Agency by the Rand Corporation (June 1980), table 4.7, p. 225.

[a] Based on a 40 g/m^3 at worst receptor.
[b] Based on a short-term, 1-hour average of 250 g/m^3.
[c] Because it is a benefit–cost study instead of a cost-effectiveness study, the Harrison comparison of the command-and-control approach with the least-cost allocation involves different benefit levels. Specifically, the benefit levels associated with the least-cost allocation are only 82 percent of those associated with the command-and-control allocation. To produce cost estimates based on more comparable benefits, as a first approximation the least-cost allocation was divided by 0.82 and the resulting number was compared with the command-and-control cost.
[d] Based on an 85 percent reduction of emissions from all sources.

aggregate emissions held constant. The possibility of saving society this much money, while not compromising air quality, is what attracts economists to alternative forms of air quality regulation.

EPA's emissions trading program. Although it began almost by accident, the Environmental Protection Agency has gradually come to embrace cost-effectiveness in regulation. In the mid-1970s the agency was grappling with a thorny legal problem: how could new sources be allowed in nonattainment areas, since such sources would exacerbate the already substandard air quality in those areas? To answer this question without cutting off growth, the EPA proposed its now well-known "offset" policy that permitted new growth in nonattainment areas provided that all new sources install pollution control equipment consistent with LAER, and offset any remaining pollution by securing even greater reductions in pollution from existing sources in the area.[51] Thus new growth could be consistent with progress toward the attainment of the national standards.

This offset policy—codified in the 1977 amendments to the Clean Air Act—provided the impetus for the development of a market in air pollution "rights," first to provide these offsets, and later to meet local emissions limits. Over time this program grew to allow the banking of pollution reduction credits if no buyer existed at the time of the reduction, as well as a "bubble" policy under which a single firm having multiple sources (stacks) could trade off added controls at one or more stacks in exchange for relaxed controls at other stacks. The purpose of all these practices, which have come to be known collectively as emissions trading, is the maintenance or improvement of air quality coupled with reductions in the cost of complying with air pollution control regulations.[52] The emissions trading program is the most ambitious attempt yet in environmental policy to capture the advantages of an incentive-based approach to environmental protection.

Encouraging though this may be, the accomplishments of emissions trading should not be exaggerated. With respect to the offset policy, the swapping of pollution rights comes into play only after new sources have installed very expensive, technology-based controls. It would make sense to consider allowing potential new sources to install less demanding technologies as long as they secure additional pollution offsets for any emissions that would result from such a downgrading of new source requirements. This would not only reduce the costs of meeting given air quality standards, but might also stimulate economic growth in those parts of the country where the closure of manufacturing plants has imposed economic hardship. Other similar restraints on emissions trading, either between firms or between separate sources at one plant, have limited the savings from the program. Nevertheless, according to one recent estimate emissions trading had resulted in cumulative cost savings of $1 to $12 billion through 1985.[53]

Serious questions have arisen about the emissions trading program. For instance, suppose Louie's Electroplating Plant is currently permitted to emit 10 tons of particulate matter per year but, for whatever reason, has historically emitted only 5 tons annually.[54] Suppose also that a new industrial facility wishing to locate in the same AQCR as Louie's proposes to offset the 3 tons of added particulates it would emit through a deal with Louie. What should be the baseline against which the trade is made? If Louie's permitted emissions limit is reduced to 7 tons per year (in exchange for the additional 3 tons from the new plant) but his actual emissions remain constant, air quality will actually worsen as a result of the deal since Louie was operating well below his limit. If Louie must reduce his actual emissions to 2 tons per year before the new plant can commence operation, then he has been denied credit for the 5 tons of emissions he has voluntarily reduced—in essence these are snatched away from him. If such a thing should happen, existing firms will be loath to voluntarily reduce emissions below permitted levels for fear the reductions would be confiscated at some future date. Thus air quality improvements would be slowed.

Emissions trading has foundered a bit on this point. Environmentalists are concerned, quite justifiably, that emissions trading will result in worse, not better, air quality because of what they call paper trades—that is, trades based on permitted rather than actual emissions. Regulated sources, on the other hand, fear the spectre of confiscation if they voluntarily reduce their emissions in anticipation of possible trades. This baseline problem must be resolved if the trading program is ever to reach its full potential.

Confident that these potential problems can be circumvented, President Bush proposed in mid-1989 to greatly expand the use of marketable permits to address both acid rain and urban smog problems. As of this writing, it was unclear what would happen to this proposal.

Congressional impediments. Unfortunately, Congress sometimes makes it difficult or impossible to reduce the cost of meeting environmental goals. The most notorious example is the 1977 change in the Clean Air Act concerning coal-fired electric power plants.[55]

In 1971 the EPA promulgated a new source performance standard that required all new coal-fired power plants to emit no more than 1.2 pounds of SO_2 per each million BTUs of electricity produced. It was expected at that time that new plants would meet this standard by incorporating into their design sophisticated equipment known euphemistically as scrubbers—actually mini-chemical plants that remove SO_2 by injecting wet limestone into the flue gases, thereby trapping (or scrubbing) the SO_2 before it escapes. Most new plants, however, chose to meet the original new source performance standards by shifting to lower-sulfur coals from southern Appalachia or western states. This saved the affected utilities, and hence their customers, a good deal of money because for most plants scrubbing would have been considerably more expensive than fuel-switching (a scrubber can account for $200 million of the $1 billion or so required to build a new 1,000-megawatt coal-fired power plant).

Letting new power plants meet the new source performance standard however best they saw fit created one problem: the future job prospects of miners of high-sulfur coal, primarily in the Midwest, were threatened. As a result, midwestern senators and congressmen—in concert with environmental groups—pushed through an amendment to the Clean Air Act in 1977 that in effect required any new coal-fired power plant not only to meet the original NSPS, but also to do so via technological means—that is, by scrubbing. According to the Congressional Budget Office, this one requirement will by the year 2000 cost rate payers $4.2 billion more *per year* in additional electricity costs than would be required to meet the same emissions standard using low-sulfur coal and allowing emissions trades between new and existing sources.[56]

How many jobs might this protectionist measure save? It is difficult to say, but one rough estimate put the number in the range of 5,000 to 10,000.[57] If correct, this estimate implies an expenditure of $400,000 to $800,000 per year *per job saved.* Since coal miners typically earn about $30,000 annually, the 1977 amendment would appear to be quite foolish—the annual incomes of the miners in question could be guaranteed for less than one-tenth the cost associated with forced scrubbing. Such an approach would seem to be far more attractive than the route Congress chose.

The vehicle emission standards that Congress has written into the Clean Air Act are also an impediment to cost-effectiveness. According to one careful review, the reductions in carbon monoxide and nitrogen oxide emissions from mobile sources which went into effect in 1981 could have been achieved for a fraction of the cost at stationary sources that emitted the same pollutants.[58] Specifically, the potential savings from a reallocation of control costs were estimated to be $240 per vehicle, implying that $2.4 billion could be saved annually (since approximately 10 million new cars are sold each year) by controlling stationary rather than mobile sources. Such potential savings suggest that there are real opportunities for emissions trading between mobile and stationary sources as well as among the latter.

Summary. Cost-effectiveness analysis of air quality regulation has been described in only the broadest terms here. Nevertheless, several observations are worth reiterating. First, the cost-effective approach avoids at least one of the difficulties that arises in benefit-cost analysis—namely, assigning dollar values to intrinsically hard-to-value outputs (lives saved, visibility improvements, and the like). Second, virtually every study of air pollution regulation has concluded that a reallocation of control effort, whether done administratively or via the creation of an emissions trading system, holds out the potential for considerable savings. Third, realizing this potential while still protecting air quality will be more difficult than is sometimes asserted. It will require agreement about the baseline for trades, may result in the (voluntary) scrapping of some pollution control equipment, and will require safeguards against the concentration at one geographic location of all or most pollutant emissions. While all these preconditions can be satisfied, it will take careful thought to do so. The possible savings are worth making the effort.

Further Economic Considerations

Many other features of the Clean Air Act have important economic implications. Although few data are available to guide our thinking about the wisdom (or lack thereof) of these provisions, a brief discussion of several of them is nevertheless in order.

Balancing in setting the NAAQSs. As stated earlier, under current legal interpretations costs may not be taken into account in setting the NAAQSs for the criteria air pollutants. In actuality, however, costs are considered whenever such standards are set, although this practice is not openly acknowledged. Rather, the decision not to adopt a standard more stringent than the one ultimately selected is justified by pointing to the uncertainty concerning the additional health protection that would be afforded by the stricter standard. Yet such uncertainty applies to virtually any standard; what is actually being admitted is that the strict standard is not worth pursuing when compared to the less stringent one, in view of the costs it would entail.

Suppose the administrator of the EPA were permitted to balance benefits and costs in setting ambient standards. How different might the individual NAAQSs look? It is difficult to say, because there have been few attempts to use benefit-cost analysis to determine the optimal standard, from an economic standpoint, for individual pollutants.[59] It is known that the costs of tightening the standards (or the savings that would result from relaxing them) are not trivial. For example, it was estimated in 1978 that meeting a 0.10 parts per million (ppm) ozone standard would cost $6.4 billion more each year in added controls on hydrocarbon and NO_x (oxides of nitrogen) emissions than the current standard of 0.12 ppm.[60] Similarly, the difference between a standard of 12 ppm for carbon monoxide rather than 9 ppm was estimated to be $0.6 to 0.8 billion per year.[61]

On the other hand, pollution control benefits can be substantial, as we have seen. Ozone, for example, is a pollutant for which the control benefits might be large. Thus the balancing of benefits and costs in setting an ozone standard might result in little or no change in the standard.[62] If the benefits of controlling other criteria pollutants were small in comparison to costs, some relaxation might result. But such a determination would depend upon a careful weighing of all the consequences of a proposed change.

It is important to keep in mind that no such new approach would work unless the EPA was allowed to make the balancing envisioned here more qualitative than quantitative. That is, while some pollution control benefits could be monetized, others probably could not—at most, a range of possible values would have to be presented. Therefore the EPA administrator's ultimate decision would—and should—reflect that person's judgments more than the dictates of a quantitative analysis.

Prevention of significant deterioration (PSD). It will be recalled from earlier discussion that PSD areas are those where air quality is better than the national standards mandate, and that while some additional pollution may be permitted in these areas, in class I and II regions the air must always be cleaner than the national standards. Class I and II areas are generally sparsely populated; that is why air quality there is so good. Note, then, the peculiar result of the approach Congress has taken: under the Clean Air Act, the highest air pollution concentrations are permitted in those areas where the most people live, while remote areas are afforded extra protection!

This is not a result one would arrive at if standard-setting were based on a calculation of where scarce pollution control resources would do the most good. The latter approach would seem to dictate that the tightest standards should exist where pollution would be the most damaging—that is, in heavily populated areas. This conclusion might not hold if the class I or II areas contained unique ecological resources or aesthetic amenities;

but even then it is hard to see why the allowable degradation should be the same in all areas, since they are not equally endowed with identical resources or amenities.

New source performance standards. The original rationale for imposing stricter controls on new sources than on existing ones appeared to be economically as well as politically sound: it is easier to incorporate air pollution control techniques in new plants than to retrofit such controls on older plants. However, the EPA has rather consistently overlooked one important economic fact in issuing new source standards since 1971: the expected life of a plant (or an automobile) is not beyond the control of the firm (or family) that owns it. By requiring strict, and therefore expensive, controls on all new sources, the EPA has created a powerful incentive to keep old plants and autos in operation longer. This in turn has delayed the improvements in air quality expected to result from the retirement of older sources and, consequently, put added pressure on the states to tighten controls on existing sources.

How great a disincentive to new plant and equipment investment have new source controls been? It is hard to say. The average age of the motor vehicle fleet in the United States has increased significantly since emissions controls were required on new cars, and at least one study has found evidence suggesting that the relationship is a causal one.[63] The differential in control between new and existing sources should also slow down the replacement of power plants, steel mills, and other affected facilities, although there are virtually no hard data demonstrating the existence, much less the magnitude, of this effect (known popularly as new source bias). A lack of data should not be taken as proof that the effect is merely imagined; indeed, this lack is probably due to the difficulty of isolating the effect of new source bias from the many other determinants of investment in plant and equipment.

No more is said here about new source bias other than to point out its potentially great importance: if environmental regulation *is* slowing down the retirement of old plants, it is not only depriving the economy of the productivity gains inherent in new technologies, but may also be at least temporarily exacerbating air quality as well. This would be doubly unfortunate.

Given the emphasis in this chapter on balancing benefits and costs, it will be hardly surprising that attention is directed to one other feature of new source standards—the requirement that they be uniform nationally.[64] The requirement might make sense if a new plant would impose roughly equal damages wherever it were located, but this is obviously not the case. An electric power plant built in a heavily populated metropolitan area would probably do much more damage than one located in a remote area. While it would be impossible for the EPA to consider a separate new source standard for every possible location, one could envision a hierarchy of standards varying from the very strict (where health or other damages would otherwise be great) to the more moderate (for remotely sited plants). Even this modest degree of tailoring would carry with it considerable cost savings.[65]

Hazardous air pollutants. Recently, one of the most controversial aspects of air pollution policy has concerned the so-called toxic or hazardous air pollutants. Although relatively little is known about them, it is worth reviewing the evidence to put the problems they may pose in perspective.

Recall that hazardous air pollutants are not nearly so ubiquitous as TSP, SO_2, or the other criteria pollutants, but are capable of causing cancer or other serious illnesses where they are encountered. Two factors have fueled the controversy surrounding their control. First, the EPA has been *very* slow to regulate hazardous air pollutants. Since 1970 only seven substances have been listed, or targeted for regulation, and some sources of these pollutants remain unregulated. In view of the much longer list of air pollutants known to be or suspected of being harmful, this record has left environmental, congressional, and other critics of the agency displeased. Second, the tragedy surrounding the accidental release of methyl isocyanate from a Union Carbide plant at Bhopal, India, heightened public awareness of the great harm that can result from airborne concentrations of harmful substances.

How serious are the risks posed by the hazardous air pollutants? As usual, the monitoring data do not provide a comprehensive picture of trends in ambient concentrations. Indeed, much less is known about the hazardous pollutants than the more common ones. However, the data that do exist suggest that ambient concentrations of a number of heavy metals have declined over the past decade.[66] So, too, have concentrations of benzo(a)pyrene, one of the most harmful of the toxic air pollutants, although the recent growth of wood-burning for home heating threatens to reverse this gain.[67]

Using a variety of approaches, a recent EPA study attempted to link ambient concentrations with estimated cancer risks by combining the pollution data with estimates of the carcinogenic potency of and human exposures to a selected group of toxic air pollutants.[68] That study, which was quite preliminary, suggests that these risks are neither trivial nor staggering. According to the report, as many as 1,500 cases of cancer *may* result each year from exposure to the group of air pollutants examined;

the actual number is likely to be smaller, perhaps much smaller, because of the conservatism inherent in risk assessment procedures. It is interesting that so-called area sources (like cars, gasoline stations, and the solvents used in many homes and workplaces) are associated in the EPA study with one-half to three-quarters of this figure. Chemical plants and other stationary sources account for the remainder.[69]

To put this estimate in some perspective, about 880,000 new cases of cancer are diagnosed each year. If the upper bound of the EPA estimate proved to be correct, the pollutants examined would account for about 0.2 percent of new cancer cases annually. Another perspective is provided by the risks associated with conventional air pollutants. According to Freeman, for example, the reduction in ambient concentrations of TSP and sulfates (a by-product of SO_2) that occurred between 1970 and 1978 has resulted in about 14,000 fewer premature deaths each year. The mortality associated with the remaining particulate and sulfur pollution could be substantially greater unless threshold concentrations exist.

It should go without saying that all these numbers are highly speculative. For instance, in several respects the EPA estimate may understate the true health risks associated with hazardous air pollutants. For one thing, only cancer risks were addressed in the report, yet a variety of other chronic and acute illnesses may result from exposure to these substances. For another, the substances examined constitute a small subset of all toxic air pollutants. While the EPA addressed what appear to be the most serious threats, further research may suggest that other pollutants pose greater risks. Also, the study did not consider the possibility of synergistic effects among pollutants.

On the other hand, the risks may have been overstated in several important respects. For example, the guidelines used by the EPA to assess the carcinogenic potency of the substances were purposely designed to err on the high side—thus the true potency of each substance is likely to be considerably less than assumed in the report. Also, the EPA assumed that individuals were exposed 24 hours a day for 70 years to prevailing outdoor concentrations of each substance. Since the average person spends from 70 to 90 percent of his time indoors, where concentrations of many of the pollutants are zero, the assumption exerts an obvious and significant upward bias on the estimate.

What is the bottom line, so to speak? Whether the risks from hazardous air pollutants are large or small, it is important to ask about the attractiveness of control options. After all, it makes sense to address even small risks if they can be quickly and inexpensively reduced. One recent study examined proposed technology-based regulations for certain sources of three serious toxic air pollutants—benzene, coke-oven emissions, and acrylonitrile.[70] The study suggests that only the coke-oven standards would reduce health risks at a cost comparable to that associated with other regulatory programs. The proposed regulations for acrylonitrile, for instance, would only make economic sense if each statistical life saved were valued at $155 million. This is nearly two orders of magnitude greater than the values typically used in benefit-cost studies. Another way to put it is to say that the resources the EPA proposed to devote to acrylonitrile controls could save many more lives if used to control other air pollutants.

Partly in response to the controversy over hazardous air pollutants, the EPA announced a new policy for these pollutants in 1985. This policy included a commitment to accelerate the listing of pollutants, increased support for state efforts to regulate toxic air pollutants that may present local but not national problems, an effort to improve emergency preparedness for serious toxic accidents, and other measures. However, in mid-1987 a court of appeals ruling found fault with the EPA's approach to hazardous air pollutant control, throwing into confusion this controversial area of regulation.[71] In response, in 1988 the Senate Committee on Environment and Public Works approved a dramatic overhaul of the way hazardous air pollutants are regulated in the United States, moving to a technology-based approach that would have called for the installation of best-available technology at most major sources. This approach failed to attract the requisite support and died in the waning days of the 100th Congress. By late 1989, though, bills embracing the technology-based approach had been reintroduced in Congress. More important, President Bush had thrown his support behind the idea as well, thus significantly increasing the likelihood that the regulation of hazardous air pollutants would be overhauled.

Interregional conflicts. Newcomers to the environmental arena may take at face value policies like those establishing stricter air quality standards in already clean areas (PSDs) or those requiring stricter emissions limits for new sources of the conventional pollutants than for older sources (NSPSs). They may be inclined to believe that the former were designed to protect national parks and other unique areas, while the latter were intended to bring about a gradual roll-over from a polluting to a clean capital stock. Some specialists (cynics?) who have studied the patterns of congressional voting on environmental issues believe otherwise. They view the PSD and NSPS policies as attempts by some regions of the United States to protect their share of business activity from encroachment by other regions—a form of domestic protectionism, if you will.

For instance, Pashigian analyzed congressional voting on the 1977 amendments to the Clean Air Act that formally established PSD as a goal of air pollution policy. After controlling for other possible determinants, he found that support for limiting allowable pollution in clean areas was stronger among congressmen representing urban and northeastern areas than among those representing soon-to-be PSD areas.[72] This evidence is consistent with (but does not prove) the hypothesis that supporters of PSD were really intent on imposing stricter environmental standards on the southern and western states that had begun to attract industrial growth away from the northeastern and midwestern states.

More recently, Crandall examined congressional voting patterns on a wide variety of environmental issues, including reclamation of strip-mined land, establishment of auto emissions standards, and the creation of a government-sponsored domestic synthetic fuels industry. Like Pashigian, he found that congressional support for the environmental position on these issues (as well as on a narrower set) was strongest among senators and congressmen representing areas where economic growth was relatively slow, that is, the so-called frostbelt. Crandall clearly believes this is no accident, arguing instead that it is an obvious response to the rapid growth of the sunbelt states in the South and West, growth which began to accelerate in the early 1970s.[73]

These analyses, and the discussion above of the hidden, job-protecting aspects of the new source standard for coal-fired electric utilities, are useful not because they prove that all environmental laws are designed to advance hidden agendas—for this they could never do—but because they illustrate the complexity of environmental policy. It is always important to look not only at the stated purpose of laws and regulations, but also at the less obvious consequences they may have. The latter may play a significant role in helping us to understand why a particular policy took the shape it did.

Acid rain. Few environmental issues—few issues of any kind —are as complicated or as interesting as that of acid rain. In some ways, it illustrates ideally the confluence of science, law, politics, economics, and ethics in public policymaking. Acid rain may also be a very serious environmental problem. So much has been written about it over the last decade or so that no chapter, much less part of a chapter, could possibly do justice to it. We must content ourselves here with only the broadest of overviews.[74]

At the most general level the acid rain problem is not too difficult to describe. Under certain photochemical conditions, emissions of sulfur dioxide and nitrogen oxides are converted to airborne particles called sulfates and nitrates, respectively. These particles can then fall to earth directly, or can interact with water vapor in the air to form mild sulfuric or nitric acids which then precipitate as rain. (Thus the acid rain problem is more correctly one of acid deposition, since the latter term encompasses both wet and dry forms.) The acids created can directly damage exposed materials, forests and agricultural products, aquatic and terrestrial ecosystems, and other valuable resources. In addition, the acidification of soils and water bodies can release aluminum and other heavy metals previously bound up in nonharmful forms. Thus acid deposition can also result in indirect damage to certain resources.

Beyond this general description, however, there is uncertainty—if not sharp disagreement—about almost all aspects of the problem, including causes and effects. For instance, although it is clear that current rainfall is sometimes quite acidic,[75] it is more difficult to ascertain whether this is a relatively new phenomenon or whether it has been with us for some time (there are few long-term monitoring records to inform this debate). Where the rainfall has become more acidic, there is frequently disagreement about the reasons: Is it due to increased SO_2 and NO_x emissions?[76] Is it due to local or distant sources or the increased use of tall stacks? Has the acidity been influenced by the removal through pollution control of certain types of particles that had acted as natural buffers to the acidic particles?

At what might be called the effects level, disagreement centers on another set of questions, including but by no means limited to the following: Has acid deposition resulted in the disappearance of fish and other aquatic life in many lakes and other water bodies? Could other causes (pesticide runoff, acid mine drainage, or conventional water pollution, for example) have contributed to this problem? Is the problem irreversible where it exists? Is acid deposition, possibly in concert with oxidant pollution, adversely affecting forest growth in the United States? Where damage is being done, can it be arrested or even reversed through the liming of water bodies, soils, and so on?

There is no shortage of answers to any of these questions. Unfortunately, they often differ. Despite these differences, however, most experts now are willing to concede that in certain parts of the United States and southern Ontario acid deposition has impaired the habitability of some water bodies. Moreover, although there is disagreement as to relative contributions, it is widely acknowledged that the long-range transport of sulfates and nitrates has added to whatever acidification may result from local emissions of SO_2 and NO_x. This is inevitable given the presence in the Midwest of many very large power plants equipped with smokestacks as tall as New York City's World Trade Towers.

How can this be happening under the Clean Air Act? This question is asked often. Yet the phenomenon of acid deposition is not necessarily inconsistent with full compliance with the requirements of the act. The Clean Air Act is designed to ensure that each air quality control region (AQCR) is in attainment with each of the NAAQSs. But these standards are based on ambient or ground-level concentrations. When SO_2 or NO_2 come out of 1,000-foot stacks, the particle by-products they often create stay airborne for days at a time and can be blown hundreds or even thousands of miles. When they are deposited from the atmosphere, they are not monitored as SO_2 or NO_2 and therefore may play no role in violating any of the air quality standards.[77] Thus it is important to understand that an acid rain problem might exist well after all parts of the country had come into compliance with each one of the NAAQSs.

Even if acid deposition did result from violations of the ambient standards, the problem still might be difficult to address because the Clean Air Act gives local authorities control over nearby sources. If Ohio (to take a hypothetical example) were lax in its enforcement of the SO_2 standard, and if some of that SO_2 were blown into Pennsylvania or New York, the latter states could petition the administrator of the EPA for protection. But it would always be difficult for the administrator (or anyone else) to determine conclusively that the problems in Pennsylvania or New York were due to Ohio's pollution rather than to local sources of SO_2. This difficulty has stood in the way of effective prohibitions on interstate pollution.

The problem of acid deposition has not gone begging for proposed solutions. Indeed, the last several sessions of Congress have witnessed a number of proposed amendments to the Clean Air Act designed to deal with the problem. These bills have differed drastically, ranging from those which would do no more than increase funding for research on acid deposition to those that would impose substantial new limitations on SO_2 (and to a lesser extent on NO_2) emissions. Those that have received the most serious consideration have been of the latter variety. While there have been some differences among the bills, they have generally called for a reduction of between 8 and 12 million tons per year in SO_2 emissions from the eastern thirty-one states (because the problem is thought to be the most serious there). This would amount to a 40 to 60 percent reduction in emissions for the affected area. However, no such bill has ever made it out of committee in the House of Representatives, nor been seriously debated on the floor of the Senate.

Why? Money, mostly, coupled with the scientific uncertainties discussed above. While estimates vary considerably, the cost associated with a 10-million-ton-per-year reduction in SO_2 emissions is on the order of $3 billion to $6 billion annually, depending on the configuration of the program. Such a cost would increase electricity rates about 3 percent on average, but in some locations the increase could be as much as 10 to 15 percent. This has proved to be a serious impediment to action, as one can imagine, especially given the uncertainty about the causes and effects of acid deposition.

The distribution of costs poses another problem for any proposed legislation. Most of the burden under any acid deposition control program would fall on a handful of states in the Ohio River Valley if the polluter pays principle is adhered to, since those states (Illinois, Indiana, Ohio, West Virginia, and Pennsylvania, among them) account for a large share of SO_2 emissions. Yet these states would reap a proportionately much smaller share of the benefits of controls: benefits would fall to the New England states and southern Ontario because they are downwind of the offending power plants and other sources. Thus a what's-in-it-for-me? attitude has characterized the reaction of one important region. Finally, if affected power plants and other sources were ordered to reduce SO_2 emissions, most would do so by shifting to lower-sulfur coals if given the choice. Once again, representatives of the high-sulfur coal states, fearing the loss of a market for their product, would be brought into the fray.

Several proposed pieces of legislation have tried to deal with all these problems at once. To soften the cost impact on the Midwest, these bills have proposed a tax on all non-nuclear electricity generation in the United States, a tax that would be used to defray the cost of SO_2 emissions reductions. To protect miners of high-sulfur coal, the bills have stipulated that some number of the dirtiest power plants would have to meet their emissions reductions by installing scrubbers—no fuel-shifting would be permitted. In trying to placate these constituencies, however, the bills have alienated western and some southern states, which have objected to being taxed to solve a problem they feel they are not a part of. These bills would also prevent them, through forced scrubbing provisions, from selling the vast reserves of low-sulfur coal upon which some sit. Moreover, the forced scrubbing provisions would make emissions reductions very inefficient in terms of cost per ton of SO_2 removed. For these and other reasons, the bills have never made it very far in Congress. President Bush's 1989 endorsement of a 10-million-tons-per-year reduction in SO_2, coupled with a limitation on annual emissions of NO_2, gave new life to the acid rain debate. This proposal, which contained no co-sharing or forced scrubbing features, was similar to proposals introduced in both houses of Congress. As of late 1989, the chances for amending the Clean Air Act to deal with acid rain had perked up considerably.

CONCLUSIONS AND POLICY RECOMMENDATIONS

Since it was notably reshaped in 1970, the Clean Air Act has become a significant feature of life in the United States. At one level, it affects the location and configuration of every new industrial facility—not to mention the time it takes to build one—as well as the conditions under which long-established plants must operate. In addition, the act has reshaped the cars we drive and the way we live, from prohibitions on leaf-burning in the fall to certain kinds of fuel-burning in the winter.

More substantively, the Clean Air Act has helped bring about cleaner air in many parts of the country, if not in absolute terms, then at least relative to what the air would have been absent the act. This in turn has affected individuals' health and the environmental amenities they enjoy, as well as the agricultural and silvicultural outputs associated with farming and forestry. While no doubt important, these benefits are of highly uncertain magnitude, despite heroic efforts to pin them down more definitively. However, these benefits have not come cheaply. Complying with the Clean Air Act has imposed significant costs on society, costs which also are uncertain. By 1990, the annual cost of complying with the Clean Air Act may be in excess of $50 billion, inevitably so if the act is amended substantially to further control SO_2 and hazardous air pollutants.

This chapter has emphasized repeatedly the difficulties that arise in comparing the benefits and costs arising from the Clean Air Act. Until both sets of estimates are refined, any attempted comparison will suffer from the apples-and-oranges problem. It follows that those favorably disposed toward the act will be able to proclaim its efficiency, while its opponents will point with glee to its apparent inefficiency. Although the range of disagreement can be—and has been—reduced considerably, the dispute will rage on.

There is relatively little disagreement about an equally important point: it is possible *right now* to substantially reduce the costs of meeting the nation's current air quality goals. This follows directly from the substantial variations in the marginal costs of controlling each of the conventional (and many hazardous) air pollutants. By reallocating control effort away from high-cost and toward low-cost sources, the total cost of pollution control can be reduced while emissions of air pollutants remain constant, or even decline. Such reallocation can be accomplished in a variety of ways, including the present command-and-control approach to air pollution control. However, the latter would require substantially more administrative and technical resources than the EPA is likely to have for some time. The same reallocation could be accomplished—indeed, *is* being accomplished—via other means, most notably through marketable emissions permits. In general, however, experiments with such incentive-based approaches have arisen idiosyncratically and have proceeded by fits and starts. It is essential that control cost reallocation be extended in a number of ways. From an economic standpoint, reallocation has the potential to make the most significant savings to date in the effort to reform environmental regulation.

Specific Recommendations

A number of suggestions for possible improvements in the Clean Air Act and the way it is administered flow from the discussion in this chapter.

1. *Introduce balancing throughout the Clean Air Act.* The CAA should be rewritten to allow the administrator of the EPA to take costs, environmental effects besides air quality, implications for energy security, and other factors into account when issuing national ambient air quality standards, new source performance standards, emissions standards for sources of hazardous air pollutants, degradation standards for clean air regions, and other ambient and source discharge standards called for in the act. However, this balancing should *not* (and probably could not) be based solely on narrow economic calculations of benefits and costs, since some values may never be expressed in dollars and cents in a way that commands broad agreement. Rather, the balancing envisioned here could be accomplished through language calling for standards to be set at levels which "are sufficient to protect public health and environmental quality to the maximum extent feasible, while taking into account the economic, environmental, and other consequences associated with these standards."

If benefits and costs are already covertly weighed in standard-setting, as suggested above, why bother to change the law, especially since in its present form it apparently gives comfort to some who would prefer to believe that absolute protection in air quality is possible? This is a hard question to answer, all the more so because any attempt to modify the Clean Air Act along the lines suggested would result in a difficult political battle. The answer is that environmental and other kinds of laws are cheapened if we continue to pretend that such considerations as costs are not germane to standard-setting, when in fact we all know they are. In other words, if we know about and implicitly accept such tradeoffs, why not sanction them in the law so that regulatory officials are forced to make tradeoffs in the open, where we can object if lives or amenities are taken too lightly in rulemaking. Indeed, other environmental statutes (the

Toxic Substances Control Act and the Federal Insecticide, Fungicide, and Rodenticide Act) *require* balancing, so the concept is not foreign.

Introducing costs and other considerations explicitly into standard-setting would have another advantage. It would eliminate the litigation that often ensues when one or another party to an environmental rulemaking gets wind of the tradeoffs that have been made. This ought to reduce the time before a standard actually takes effect and is enforced, and should also reduce the uncertainty that attends many EPA decisions while all parties await judges' sometimes puzzling rulings. The balancing envisioned here would not be satisfied by an affordability criterion in the act. A particular standard might be eminently affordable but nonetheless downright silly if it accomplished little of value to the citizenry. Thus the affordability requirement currently a part of the NSPS provisions of the act does not constitute the kind of authority to consider costs that is called for here.

As envisioned here, balancing would be practiced for each ambient air quality standard to be set, but only because there are relatively few of them. In the case of discharge standards for new sources (NSPS, BACT, or LAER), it would be impossible to consider a separate standard for each and every new source, in each possible location, based on a comparison of benefits and costs. Rather, the spirit of balancing would be satisfied by establishing sets of new source standards depending on location. Thus a new source in a heavily populated area or one where there are delicate ecosystems might be required to meet strict emissions limits. A new source locating in a less populated or less ecologically sensitive area would still face emissions limits, but they might be less demanding.

2. *Reduce new source bias*. As it has evolved, regulation under the Clean Air Act requires much greater control efforts from new sources than from existing ones. This disparity must be narrowed. It is an institutional impediment to equalizing marginal control costs across sources and thus stands in the way of reducing the costs of meeting current air quality goals.

The disparity can also discourage the construction of new plants and the retirement of old ones. It is even anti-environmental in that it delays the date at which older, more polluting facilities close down. The same principle applies to cars. The CAA presently places the major burden of pollution control on carmakers, who accomplish this task through catalytic converters and other hardware. Yet it is often much less expensive to reduce emissions through tune-ups, proper fueling of cars, and other practices that should be the responsibility of the owner rather than the manufacturer. These actions are seldom taken because many states do not have vehicle inspection programs, and those that do often operate them in a pro forma way. If more of the burden for automobile pollution control fell on owners, it would be possible to reduce somewhat the controls imposed on new vehicles. This would accelerate the roll-over of the stock of older, highly polluting vehicles.

In areas that continue to struggle to meet the ozone or carbon monoxide standards, it will be necessary to both reduce emissions from older vehicles and adhere to strict standards for new vehicles. In these areas the tradeoffs envisioned here may not be feasible.

3. *Expand emissions trading.* One way to reduce new source bias is through expanded emissions trading between new and existing sources. That is, a new source should be permitted to emit more pollution than is currently allowed if—and only if—it can secure even greater emissions reductions from existing sources in the area in which it locates. This principle should apply regardless of whether the source is locating in a nonattainment, PSD, or other area.

Expanded emissions trading should not be confined to new sources, even though its most important application might be at these sources. Rather, trading should be expanded among existing sources as well as among individual emissions points within one facility. In other words, the bubble policy should be broadened wherever possible and should be actively encouraged by the EPA and by Congress. If an acid rain control program is eventually adopted, it should include provisions for widespread emissions trading among affected sources, much as President Bush proposed in July 1989. The substantial costs of any significant acid rain program could be reduced through such trading.

Important as this point is, it is equally important that any expansion of the trading program be carefully overseen. Incentive-based approaches to environmental policy will never succeed if they are ill-concealed attempts to circumvent existing pollution control requirements.

4. *Improve the inventory of data on current emissions.* No system of exchange, in financial or pollution control markets, can succeed if property rights are not clearly delineated. Thus it is extremely important that the current data on who's discharging what be improved. Continuous monitoring of emissions must gradually replace the practice of estimating emissions, at the very least for large point sources. This would not only help form the basis for a system of marketable pollution permits, but would also assist in documenting violations of existing standards. To put the matter bluntly, it seems strange to spend as much for pollution control

as the United States now does while at the same time estimating whether or not regulated sources are in compliance with their standards.

5. *Expand and improve ambient monitoring.* Because the major emphasis in this chapter has been on policy, and because a separate chapter is devoted to monitoring, little has been said here about the air pollution monitoring network. Nevertheless, it will be difficult if not impossible to make good policy without having timely, reliable data on ambient environmental conditions. How else can we know whether current policies are working, or whether new policies should be tried? How else can we know whether rural air pollution problems are growing worse even as urban problems appear generally to be on the wane? How else can we determine the magnitude of the hazardous air pollutant problem? There is simply no substitute for air quality data in judging the effectiveness of air quality programs.

Three types of improvements suggest themselves. First, more air pollution monitors must be sited in rural areas. This will help establish background concentrations, shed much-needed light on the possible seriousness of air pollution threats to agriculture and silviculture, and indicate the extent of long-range transport of pollutants from urban or industrialized areas to less populated or developed areas. Second, a monitoring network for hazardous air pollutants must be created. It makes little sense to redirect air pollution policy toward these pollutants (as some would do) in the absence of information about the extent of the problems they pose. Third, monitoring must be reoriented toward the right problems. For instance, until recently particulates were measured by TSP (which includes very large particles) even though there is broad agreement that only the smaller particles pose a serious threat to health. (This is why in 1987 the EPA changed the basis of the particulate standard to those particles less than 10 microns in diameter.) Thus we should quickly reorient our monitoring network to focus on the so-called fine or inhalable particulates, particularly the acidic sulfate particles that are the most serious threat to health.

None of these recommendations will be simple to put in place. They will require money, political capital, or both. Nor will these recommendations solve the problems that arise in air pollution policy; any substantial effort to reallocate resources in modern society will encounter difficulties. But each of the recommendations should make a modest contribution to the establishment of sensible goals for air quality and the development of effective and efficient means of attaining these goals.

NOTES

1. For a discussion of the Reagan administration's regulatory program, see George C. Eads and Michael Fix, *Relief or Reform? Reagan's Regulatory Dilemma* (Washington, D.C., Urban Institute Press, 1984).

2. This section draws heavily on Arthur C. Stern, "History of Air Pollution Legislation in the United States," *Journal of the Air Pollution Control Association* vol. 32, no. 1 (January 1982) pp. 44–61.

3. J. Clarence Davies, *The Politics of Pollution* (New York, Pegasus Press, 1970).

4. Stern, "History of Air Pollution Legislation," p. 44.

5. For an interesting history of motor vehicle pollution control, see Lawrence J. White, *The Regulation of Air Pollutant Emissions from Motor Vehicles* (Washington, D.C., American Enterprise Institute, 1982).

6. See Robert D. Friedman, *Sensitive Populations and Environmental Standards* (Washington, D.C., The Conservation Foundation, 1981).

7. Even this step would not reduce risks to zero, since there are natural sources for virtually all the common air pollutants, including windblown dust, volcanoes, sea spray, the decomposition of organic matter, and many others.

8. See National Academy of Sciences, *On Prevention of Significant Deterioration* (Washington, D.C., National Academy Press, 1981).

9. See White, *Regulation of Air Pollutant Emissions,* pp. 15–19, for an extended discussion of these standards.

10. New sources wishing to locate in nonattainment areas are also required to offset any emissions remaining after LAER by securing reductions from sources already in the area.

11. Monitoring deficiencies are discussed more extensively in Robert W. Crandall and Paul R. Portney, "Environmental Policy," in Paul R. Portney, ed., *Natural Resources and the Environment: The Reagan Approach* (Washington, D.C., Urban Institute, 1984) pp. 47–81.

12. While unfortunate, this is not as crucial an impediment as it may seem to understanding the effects of the Clean Air Act. Many of the controls called for in the act were not put in place until several years (and a number of lawsuits) had passed. Thus even if quite effective, the controls embodied in the act probably would not have yielded significant improvements until 1973 or 1974. For this reason, the mid-1970s is not a bad place to begin.

13. Information provided by the Environmental Protection Agency (EPA).

14. The EPA's Office of Research and Development is required to review periodically the clinical, toxicological, and epidemiological evidence concerning the health effects associated with the criteria air pollutants. These reviews are issued in the form of huge documents which purport to summarize and interpret the studies. A cursory reading of any of these reports—called criteria documents—will illustrate the ambiguities inherent in the studies' findings.

15. See National Academy of Sciences, *Epidemiology and Air Pollution* (Washington, D.C., National Academy Press, 1985).

16. Even a shuffling of plant locations like this, however, would have the benefit of reducing the number of people exposed to pollution. Although total pollution emissions may remain constant, air pollution damages could fall.

17. Given constant pollutant emissions in a particular area, ambient ozone concentrations can vary significantly, depending on the amount of sunlight and the strength of the

prevailing winds. Similarly, particulate concentrations will be lower during rainy periods than dry ones, even if particulate emissions are constant.

18. Ivy E. Broder, "Ambient Particulate Levels and Capital Expenditures: An Empirical Analysis," undated paper, Department of Economics, American University; Paul W. MacAvoy, "The Record of the Environmental Protection Agency in Controlling Industrial Air Pollution," in R. L. Gordon, H. D. Jacoby, and M. B. Zimmerman, eds., *Energy Markets and Regulation* (Cambridge, Mass., Ballinger, 1987) pp. 107–137.

19. MacAvoy does make one pass at explaining ambient concentrations of particulate matter; see MacAvoy, "Record of the Environmental Protection Agency," pp. 73–76.

20. See Paul R. Portney, "The Macroeconomic Impacts of Federal Environmental Regulation," *Natural Resources Journal* vol. 21 (July 1981) pp. 459–488.

21. Environmental Protection Agency, "1980 Ambient Assessment—Air Portion," Report no. EPA-4501 4-81-014 (February 1981) pp. 2–1 through 2–6.

22. On the emissions record of in-use vehicles over time, see White, *Regulation of Air Pollutant Emissions*, pp. 29–36, 56.

23. For a discussion of benefit-cost analysis in general, see Edward N. Gramlich, *Benefit-Cost Analysis of Government Programs* (Englewood Cliffs, N.J., Prentice-Hall, 1981); for a theoretical treatment of environmental benefit estimation, see A. Myrick Freeman III, *The Benefits of Environmental Improvement: Theory and Practice* (Baltimore, The Johns Hopkins University Press for Resources for the Future, 1979); for a clear and nontechnical discussion of recent studies of environmental benefits, see Allen V. Kneese, *Measuring the Benefits of Clean Air and Water* (Washington, D.C., Resources for the Future, 1984).

24. See Portney, "Macroeconomic Impacts."

25. See Jon C. Sonstelie and Paul R. Portney, "Truth or Consequences: Cost Revelation and Regulation," *Journal of Policy Analysis and Management* vol. 2, no. 2 (1983) pp. 280–285.

26. A. Myrick Freeman III, "The Benefits of Air and Water Pollution Control: A Review and Synthesis," Report to the Council on Environmental Quality, Executive Office of the President, April 1980; A. Myrick Freeman III, *Air and Water Pollution Control: A Benefit-Cost Assessment* (New York, Wiley, 1982).

27. Freeman does cite some evidence, however, suggesting that for particulates and sulfur dioxide, the improvements in air quality between 1970 and 1978 were due to emissions reductions resulting from the Clean Air Act. See Freeman, *Air and Water Pollution Control*, p. 31.

28. Cold as they may seem, these estimates are not arbitrary assignments of value to precious lives. Rather, they are all based on individuals' revealed willingness to trade off *slight* increases in the risk of death (usually in the workplace) for generally *small* increases in money (usually in the form of wages). These studies do not imply that any individual would be indifferent to the choice between certain death and a large amount of money.

29. All monetary sums in this chapter are expressed in 1984 dollars, unless otherwise noted.

30. Environmental Protection Agency, "National Air Quality and Emissions Trends Report, 1987," Document no. EPA-450/4-89-01 (March 1989) pp. 40–68.

31. White, *Regulation of Air Pollutant Emissions*, p. 55.

32. Some evidence suggests that the regulation of hazardous air pollutants under Section 112 of the Clean Air Act of 1970 may not give rise to substantial benefits. See John A. Haigh, David Harrison, Jr., and Albert L. Nichols, "Benefit-Cost Analysis of Environmental

Regulation: Case Studies of Hazardous Air Pollutants," *Harvard Environmental Law Review* vol. 8, no. 2 (1984) pp. 395–434.

33. See, for example, R. M. Adams, S. A. Hamilton, and B. A. McCarl, "An Assessment of the Economic Effects of Ozone on U.S. Agriculture," *Journal of the Air Pollution Control Association* vol. 35, no. 9 (September 1985) pp. 938–943.

34. See G. S. Tolley and coauthors, "Establishing and Valuing the Effects of Improved Visibility in the Eastern United States," draft report, Department of Economics, University of Chicago, to Environmental Protection Agency, 1985.

35. It should be noted, however, that at least one study has found associations between air pollution and elevated mortality rates even among the very young. See Lester B. Lave and Eugene P. Seskin, *Air Pollution and Human Health* (Baltimore, The Johns Hopkins University Press, 1977).

36. Public Interest Economics Foundation, "The Aggregate Benefits of Air Pollution Control," prepared for the Office of Policy Analysis, Environmental Protection Agency, 1984.

37. Mathtech, Inc., "Benefits and Net Benefit Analysis of Alternative National Ambient Air Quality Standards for Particulate Matter," report for the Economic Analysis Branch, Office of Air Quality Planning and Standards, Environmental Protection Agency, under contract no. 68-02-3826 (March 1983).

38. Specifically, the estimates assume implementation of a standard limiting annual average ambient concentrations of particulate matter of less than 10 microns in size to no more than 70 micrograms per cubic meter ($\mu g/m^3$), and average 24-hour concentrations to no more than 250 $\mu g/m^3$.

39. This approach was selected as one way to deal with existing disagreements over the adverse effects of particulate matter. Those who have no confidence in studies linking particulate concentrations to acute morbidity or household cleaning will prefer scenario A. Others, particularly those convinced by studies finding large adverse effects on health and materials, will feel more comfortable with scenarios D–F, since the latter reflect such studies.

40. Office of Policy Analysis, Environmental Protection Agency, "Costs and Benefits of Reducing Lead in Gasoline," Report no. EPA-230-05-85-006 (February 1985).

41. Net benefits were also presented without blood-pressure effects, because the studies on which these benefits were based were quite new and therefore perhaps speculative.

42. It is with considerable humility that this catch-as-catch-can approach is described, since I was personally responsible for the CEQ's estimates in its 1979 and 1980 annual reports to the president.

43. Temple, Barker, and Sloan, Inc., *Environmental Regulations and the Electric Utility Industry*, report to the Office of Policy Analysis, Environmental Protection Agency (December 1982).

44. Robert W. Crandall, Howard K. Gruenspecht, Theodore E. Keeler, and Lester B. Lave, *Regulating the Automobile* (Washington, D.C., Brookings Institution, 1986).

45. See White, *Regulation of Air Pollutant Emissions*, pp. 59–66.

46. Frederick Kappler and Gary L. Rutledge, "Expenditures for Abating Pollutant Emissions from Motor Vehicles, 1968–84," *Survey of Current Business* (July 1985) pp. 29–35.

47. See Michael Hazilla and Raymond J. Kopp, "The Social Cost of Environmental Quality Regulations: A General Equilibrium Analysis," Discussion Paper QE86-02, Resources for the Future (August 1986).

48. This does not imply that environmental expenditures are not productive, only that the benefits often do not show up in traditional GNP accounts.

49. T. H. Tietenberg, *Emissions Trading: An Exercise in Reforming Pollution Policy* (Washington, D.C., Resources for the Future, 1985).

50. Ibid., p. 48.

51. For an excellent discussion of the origins and development of this program, see Richard A. Liroff, *Air Pollution Offsets* (Washington, D.C., The Conservation Foundation, 1980).

52. See Richard A. Liroff, "The Bubble Policy and Emissions Trading: The Toil and Trouble of Regulatory Reform," draft, The Conservation Foundation, 1985.

53. Robert W. Hahn and Gordon L. Hester, "Where Did All the Markets Go? An Analysis of EPA's Emissions Trading Program," *Yale Journal on Regulation* vol. 6, no. 1 (Winter 1989).

54. This is not as farfetched as it may seem. See Winston Harrington, *The Regulatory Approach to Air Quality Management: A Case Study of New Mexico* (Washington, D.C., Resources for the Future, 1981).

55. The history of this debacle is spelled out in Bruce A. Ackerman and William T. Hassler, *Clean Coal/Dirty Air* (New Haven, Yale University Press, 1981).

56. Congressional Budget Office, "The Clean Air Act, the Electric Utilities, and the Coal Market" (Washington, D.C., Government Printing Office, 1982). For this chapter, CBO estimates were converted to 1984 dollars using consumer price indexes.

57. Paul R. Portney, "How *Not* to Create a Job," *Regulation* (November/December, 1982) pp. 35–38.

58. White, *Regulation of Air Pollutant Emissions,* pp. 84–88.

59. For one example of such an attempt, see Wallace E. Oates, Paul R. Portney, and Albert M. McGartland, "The *Net* Benefits of Incentive-Based Regulation: A Case Study of Environmental Standard-Setting," *American Economic Review* (forthcoming).

60. Lawrence J. White, *Reforming Regulation: Processes and Problems* (Englewood Cliffs, N.J., Prentice-Hall, 1981) pp. 61–64. 1978 dollars have been converted to 1984 dollars using consumer price index.

61. Council on Wage and Price Stability, Executive Office of the President, "Environmental Protection Agency Proposed National Ambient Air Quality Standards for Carbon Monoxide," report submitted to the EPA, November 24, 1980.

62. We would not expect balancing to lead to a tightening of controls unless that process called attention to studies that were ignored in setting the current standards.

63. Howard K. Gruenspecht, "Differentiated Regulation: The Case of Auto Emissions Standards," *American Economic Review* vol. 72, no. 2 (May 1982) pp. 328–331.

64. This does not pertain to nonattainment areas where new sources are supposed to meet even stricter standards (LAER areas).

65. See David Harrison, Jr., and Albert L. Nichols, "Benefit-Based Flexibility in Environmental Regulation," Kennedy School of Government, Harvard University, Working Paper no. E-83-06 (April 1983).

66. Information provided by the Environmental Protection Agency.

67. "Estimation of Cancer Incidence Cases and Rates for Selected Toxic Air Pollutants Using Ambient Air Pollution Data, 1970 Versus 1980," report prepared by the Office of Air Quality Planning and Standards, Environmental Protection Agency (April 1985).

68. "The Air Toxics Problem in the United States: An Analysis of Cancer Risks for Selected Pollutants," report prepared by the Office of Air and Radiation and the Office of Policy Analysis, Environmental Protection Agency (May 1985).

69. Ibid, p. 82.

70. Haigh, Harrison, and Nichols, "Benefit-Cost Analysis of Environmental Regulation."

71. *Natural Resources Defense Council Inc., v. U.S.E.P.A.*, 458 F 2nd 1146 (D. C. Cir. 1987).

72. Peter B. Pashigian, "Environmental Regulation: Whose Self Interests Are Being Protected?" *Economic Inquiry* (October 1985) pp. 551–584.

73. Robert W. Crandall, *Controlling Industrial Pollution* (Washington, D.C., Brookings Institution, 1983) pp. 110–130.

74. Perhaps the best overview is Office of Technology Assessment, U.S. Congress, *Acid Rain and Transported Air Pollutants: Implications for Public Policy*, OTA-O-204 (Washington, D.C., June 1984). See also Winston Harrington, *Acid Rain: Science and Policy* (Washington, D.C., Resources for the Future, 1989).

75. Acidity is measured on what is known as the pH scale, where a pH of 14 is the most alkaline and pH of 1 is the most acid. "Normal" rainfall would be slightly acid (a pH of about 5.7) due to the presence of mild carbonic acid resulting from airborne CO_2. Virtually the entire Northeast and Midwest were exposed to rainfall with an average pH of 4.2 to 4.5 in 1980, and in parts of New York, Ohio, West Virginia, Kentucky, and New Jersey the average pH was less than 4.2 (see Office of Technology Assessment, *Acid Rain and Transported Air Pollutants*, p.6). Because the pH scale is logarithmic, this means that rainfall in these areas was at least 10 times and as much as 60 to 70 times more acidic than normal.

76. One of the peculiarities of the debate is that it has arisen during a period when nationwide emissions of SO_2 are estimated to have fallen by nearly 20 percent (see table 3–4).

77. It is possible that the resulting sulfate or nitrate particles could contribute to the violation of the current TSP standard in the receiving locations. This problem has received virtually no attention in the debate over acid rain, however.

four

Water Pollution Policy

A. Myrick Freeman III*

INTRODUCTION

The water in our lakes, rivers, and streams supports a wide range of uses. Water can be withdrawn for drinking and other domestic purposes, for industrial processes, or for irrigation. It can support fish populations that are the basis of commercial exploitation and recreational fishing. It can be used for boating and swimming. And it can be used to flush away the wastes from factories and municipal sewers. Most of these uses are to varying degrees dependent on the quality of the water. Yet the use of a water body as a waste receptor can seriously degrade water quality and impair or even preclude other uses.

A Ralph Nader Task Force Report, *Water Wasteland*, published in 1971, helped to dramatize the poor state of some of our water bodies and may have helped to spur Congress to overhaul our federal water pollution control laws.[1] Congress enacted major revisions in federal water pollution law in 1972. These revisions are known as the Federal Water Pollution Control Act Amendments of 1972 (FWPCA-72). This law established goals for water quality and called for the elimination of all discharges of pollutants into navigable waters by 1985. The law also established a National Commission on Water Quality to study the technical, economic, social, and environmental aspects of achieving some of the specific provisions of the act. The commission issued its required report in 1976; this report helped to guide Congress in revisions to the 1972 law that were adopted in 1977.[2] The amended act is now known as the Clean Water Act (CWA).

*The author is indebted to Paul R. Portney for a number of helpful comments and suggestions during the preparation of this chapter.

For several reasons this is a good time to take a new look at our federal water pollution control policies. First, we now have more than a decade of experience with the Clean Water Act since the midcourse corrections of 1977. Second, the target dates in the act for achieving the water quality goals and the elimination of pollution discharges have now passed; thus it is important to know what progress has been made toward achieving these goals and whether the goals themselves provide a useful or meaningful basis for federal water pollution control policy. And third, since Congress reauthorized the act in 1987, we can examine the extent to which the changes made during the reauthorization deal effectively with the problems that have emerged over the past fifteen years.

The next section of this chapter provides a brief review of the history and evolution of federal water pollution control policy, highlighting some of the changes and concerns which have guided policy in the direction of greater federal responsibility for goal-setting, implementation, and financing. In the third section we discuss those key features of FWPCA-72 that define the present approach to water pollution control policy. We also discuss the revisions to this framework which have been adopted through amendments in 1977, 1981, and 1987. In the fourth section we review, using several measures of performance, what has actually been accomplished in controlling discharges and improving the quality of our nation's waters, to the extent that these can be known.

We then turn to an economic assessment of water pollution control policy, including a comparison of benefits and costs, and the problem of cost-effectiveness. Some of these economic principles are then applied to evaluate the past performance and possible modifications of the federal program for subsidizing construction of municipal sewage treatment plants and the evolving federal approach to the control of non-point source pollution.

THE HISTORY AND EVOLUTION OF WATER POLLUTION POLICY

The history of the federal government's involvement in controlling water pollution begins, in a way, with the Refuse Act of 1899. This act, the purpose of which was to prevent impediments to navigation, stated that "it shall not be lawful to throw, discharge, or deposit . . . any refuse matter of any kind . . . into any navigable water of the United States. . . ."[3] The act is a useful reminder that at one time the sheer physical volume of waste discharges from some industries—for example, sludge, fiber, and sawdust from paper mills and sawmills—threatened to block some rivers and channels. However, the Refuse Act, as administered by the U.S. Army Corps of Engineers, had no impact on most industrial and municipal sources of water pollution.[4]

The first federal legislation to deal explicitly with conventional forms of water pollution was the Water Pollution Control Act of 1948.[5] Summaries of the key features of this and subsequent federal laws dealing with water pollution control policy are provided in table 4–1. In the 1948 law, Congress recognized a national interest in controlling water pollution by authorizing the federal government to engage in research, investigation, and surveys dealing with water pollution problems—activities not unlike those called for in the first federal legislation on air pollution control (see chapter 3). The Water Pollution Control Act also authorized the federal government to make loans to municipalities for construction of municipal sewage treatment facilities. Again as in the air pollution case, there was no federal authority to establish water quality standards, limit discharges, or engage in any form of enforcement.

A key feature of the subsequent Water Pollution Control Act Amendments of 1956 was the establishment of a federal program of direct grants to municipalities to share in the costs of constructing sewage treatment facilities. The federal share of construction costs was set at 55 percent. The 1956 amendments also sanctioned a goal-oriented approach to pollution control policy by authorizing states to establish criteria for determining desirable levels of water quality. This act also created a federal role in establishing and enforcing clean-up requirements on individual dischargers through the mechanism of the "enforcement conference." If a serious water pollution problem were recognized in interstate waters, the responsible federal agency (the Public Health Service in the Department of Health, Education, and Welfare) could, at its discretion, convene a meeting of state and local officials, major polluters, and other interested parties to make recommendations as to who should clean up and by how much. However, the combination of discretionary action on the part of the federal agency and reliance on consensus and volunteerism on the part of participants in the conferences meant that this act made little contribution to the establishment and enforcement of meaningful pollution control requirements on individual dischargers.

The Water Quality Act of 1965 was the first federal law to mandate state actions with respect to water pollution control policy. This act required that states establish ambient water quality standards for interstate water bodies and develop implementation plans for controlling pollution sufficiently to meet these standards. Implementation plans were to call for specified reductions in pollution discharges from individual sources.

Table 4–1. Federal Water Pollution Control Laws

Title and year of enactment	Key provisions
The Refuse Act, 1899	*Goals:* protection of navigation *Means:* barred discharge or deposit of refuse matter in navigable waters without permit *Federal vs. state responsibility:* federal permits and enforcement *Financing of municipal sewage treatment:* none
Water Pollution Control Act, 1948	*Goals:* encouragement of water pollution control *Means:* authority for federal research and investigation *Federal vs. state responsibility:* left to state and local governments *Financing of municipal sewage treatment:* authorized federal loans for construction, but no funds were appropriated
Water Pollution Control Act Amendments, 1956	*Goals:* authorized states to establish water quality criteria *Means:* federally sponsored enforcement conferences to negotiate clean-up plans *Federal vs. state responsibility:* federal discretionary responsibility to initiate enforcement conferences for interstate waters *Financing of municipal sewage treatment:* authorized federal grants to cover up to 55% of construction costs
Water Quality Act, 1965	*Goals:* attainment of ambient water quality standards required to be established by states *Means:* state-established implementation plans placing limits on discharges from individual sources *Federal vs. state responsibility:* state responsibility for setting standards, developing implementation plans, and enforcement; federal oversight through approval and strengthened enforcement conference procedures *Financing of municipal sewage treatment:* no significant change

Federal Water Pollution Control Act, 1972	*Goals:* fishable and swimmable waters *Means:* enforcement of technology-based effluent standards on individual dischargers *Federal vs. state responsibility:* federal responsibility for establishing effluent limits for categories of sources, and for issuing and enforcing terms of permits to individual dischargers; state option to take over responsibility for permits and enforcement *Financing of municipal sewage treatment:* federal share increased to 75% and total authorization substantially increased ($18 billion over 3 years)
Clean Water Act, 1977	*Goals:* postponement of some deadlines established in the 1972 act; increased control of toxic pollutants *Means:* no significant changes *Federal vs. state responsibility:* no significant changes *Financing of municipal sewage treatment:* no significant changes (authorizations for an additional $25.5 billion in federal grants over 6 years)
Municipal Wastewater Treatment Construction Grant Amendments, 1981	*Financing of municipal sewage treatment:* reduced federal share to 55%, changed allocation priorities, and lowered authorizations to $2.4 billion per year for 4 years
Water Quality Act, 1987	*Goals:* further postponement of deadlines for technology-based effluent standards *Financing of municipal sewage treatment:* transition from federal grants to contributions to state revolving loan funds

Sources: Adapted from Allen V. Kneese and Charles L. Schultze, *Pollution, Prices, and Public Policy* (Washington, D.C., Brookings Institution, 1975) p. 30, table 3.1; and T. H. Tietenberg, *Environmental and Natural Resource Economics* (Glenview, Ill., Scott, Foresman and Co., 1984) p. 51, table 15.1, with additional information provided by the author.

States also had primary responsibility for any enforcement action against individual sources not in compliance with implementation plans. The 1965 act also provided for federal oversight of action taken by states by requiring federal approval of water quality standards and implementation plans, and by providing a somewhat strengthened enforcement conference procedure.

Under the standard-setting provisions of the 1965 law, each state was required to draw up minimum water quality standards for those portions of interstate waters within its borders. These standards could vary for different bodies of water. Although the law was not specific on this point, the standard-setting requirement gave states the opportunity to weigh, at least implicitly, the benefits and costs of attaining different levels of water quality. After setting water quality standards, states were to determine the maximum discharges of various pollutants consistent with meeting these standards. The total allowable discharges would then be divided among the major sources of discharges on some basis. Dischargers would receive permits specifying the quantity of wastes they could legally discharge. Detecting violations of permits, taking offending dischargers to court, and imposing fines where necessary were also the responsibilities of the states.

For three reasons this system of standard-setting and enforcement by states came to be viewed by many as ineffective and unworkable.[6] The first was the difficulty in determining how much each individual source must cut back its discharges in order for the predetermined water quality standards to be met along a whole river basin. In principle a water quality model could have been developed for each river basin and used to determine that set of discharge reductions which would meet the water quality standards at minimum aggregate pollution control cost.[7] But in practice the development and use of such models is costly and time-consuming, and was beyond the analytical capabilities of many state pollution control agencies. Hence in practice many state agencies applied simple rules of thumb such as requiring secondary treatment or its equivalent for pollutant discharges from all sources.[8]

A second reason involved the enforcement of individual discharge standards when water quality violations were detected. The problem was in establishing a legally satisfactory way by which one or more dischargers could be held responsible for causing a violation of the standards. In other words, some means had to be established to determine who was at fault when water quality fell below the minimum acceptable level.

The third reason for viewing the system as unworkable concerned the problem of leaving the responsibility for enforcement primarily with the individual states. States varied enormously in their commitment to pollution control objectives, in the talent of their personnel, in the resources they could make available for implementation, monitoring, and enforcement, and in their willingness to resist the temptation to compete for new industry by offering "friendly" regulatory environments.

Responding to these concerns, Congress passed a major revision to federal water pollution control policy in 1972. The Federal Water Pollution Control Act of 1972 established new goals and standards for water quality, set deadlines for clean-up actions, and created new procedures and mechanisms for regulation and enforcement. These revisions represented a major departure from the earlier approach in three main respects: goals, methods, and federal responsibility. (For a detailed discussion of FWPCA-72 and amendments to it, see the next section of this chapter.)

As we have seen, FWPCA-72 established as a national goal of water pollution policy the elimination of all discharge of pollutants into navigable waters by 1985 and the attainment of fishable and swimmable waters by July 1, 1983. The means selected for achieving this goal would be a system of technology-based effluent standards to be established by the EPA. These standards would define the maximum quantities of pollutants each source would be allowed to discharge. By basing effluent standards strictly on technological factors rather than water quality objectives, the 1972 amendments did away with the need for regulators to estimate the assimilative capacity of water bodies and the relationship between individual discharges and water quality. The FWPCA-72 revisions called for the application of the same effluent standards to all dischargers within classes and categories of industries, rather than a plant-by-plant determination of acceptable discharges based on water quality considerations.

The third major departure of FWPCA-72 from past policy was that, at least initially, the major responsibility for issuing permits to dischargers was shifted to the federal government, specifically to the Environmental Protection Agency. All dischargers would have to hold permits and comply with their terms, which were to embody the technology-based effluent standards to be set by the EPA. The law called for the EPA to turn over the responsibility for issuing permits to individual states when these states met certain conditions.

While making major changes in the objectives and approach to enforcement, FWPCA-1972 also continued and strengthened the program of federal subsidies to cities and towns for treating municipal wastes. The 1972 amendments raised the federal share of treatment plant construction costs to 75 percent, and substantially increased the amounts authorized to be spent. The authorizations for the fiscal years 1973 through 1975 totaled $18 billion. These subsidies were as high as $7 billion in fiscal year 1975, and have been running at close to $2.5 billion per year in recent years.

By 1977 Congress wished to make further changes in policy. The Clean Water Act of 1977 represented a midcourse correction to the path outlined in the 1972 act. There were two major changes embodied in the 1977 act. The first was the postponement of several deadlines established in the 1972 act for compliance with effluent standards by individual dischargers. The second involved making a clearer distinction between the so-called conventional pollutants (organic matter and suspended solids, for example) that were the focus of FWPCA-1972 and the so-called toxic water pollutants about which concern had been mounting. The 1977 act also established new procedures and deadlines for determining effluent limitations for toxic pollutants and ensuring compliance with these limitations by individual dischargers.

The most recent changes in the law were made as part of the reauthorization of the Clean Water Act in early 1987. The Water Quality Act of 1987 was passed over President Reagan's veto as the first major piece of business of the 100th Congress. The president's principal objection was to the continued high level of federal aid for the construction of municipal sewage treatment facilities. Congress authorized $9.6 billion in construction grants through fiscal year 1990, and an additional $8.4 billion through 1994 in federal contributions to state revolving loan funds to support continued construction after the end of the federal grant program. The 1987 act also further postponed the deadlines for compliance with effluent standards and established new requirements for states to develop and implement programs for controlling so-called non-point sources of pollution. Such sources include runoff from cultivated agricultural land, from silvicultural activities, and from urban areas.

Three major trends in the evolution of a national policy toward water pollution control are evident in this brief review of federal laws. The first is the shift in the focus of responsibility for setting goals, implementing policy, and enforcing pollution control requirements from the state and local levels to the federal government. Before 1948, there was no federal role in water pollution control policy except for controlling impediments to navigation. It was not until 1956 that the federal government assumed any responsibility for enforcement of pollution control requirements. But by 1972 the federal government had assumed the responsibility for setting pollution control objectives and assumed primary responsibility for implementation and enforcement.

The second trend is the shift in financial responsibility for constructing municipal sewage treatment plants from state and local governments to a significant federal cost-sharing program. Municipalities have been constructing sewage treatment works throughout this century, with no help from the federal government. In 1956 the first federal construction grant program was authorized. And by 1972 the federal share of the cost had been increased to 75 percent of capital costs; authorizations were generous enough so that every municipal project which met minimum requirements could expect to receive its federal grant sooner or later. This trend was reversed somewhat by amendments in 1981 and 1987, but a significant federal role is assured at least through 1994.

The third major trend has been in the nature of the objectives of water pollution control policy. In the early years of this century the major justification for the construction of municipal sewage treatment facilities was to reduce the flow of disease-carrying human wastes into rivers and streams, where they could pose threats to human health through contamination of drinking water, shellfish beds, and other resources. But beginning at least with the 1965 act, federal policy reflected an increased concern with the protection of so-called in-stream uses of water bodies (such as swimming and boating) as well as ecological values. The water quality standards that were to be adopted under the 1965 Water Quality Act were meant to assure water of sufficient quality to support various existing or desired uses of water bodies including boating, fishing, and swimming. The 1972 amendments made the achievement of fishable and swimmable water a national policy objective. And the Clean Water Act of 1977 gave increasing emphasis to "the protection and propagation of a balanced population of shellfish, fish, and wildlife in the establishment of effluent limitations."[9]

THE FEDERAL WATER POLLUTION CONTROL ACT OF 1972 AND ITS AMENDMENTS

Because of the importance of the changes brought about by the Federal Water Pollution Control Act of 1972, and because subsequent legislation has not significantly altered the basic goals and means of federal policy, we will examine here in more detail the key features of the act and its amendments. Since FWPCA-72 takes quite different approaches toward point source pollution and non-point source pollution, this section concludes with a discussion of those provisions of the act relating to non-point sources.

Goals

The first sentence of FWPCA-72 states that the goal of the act is to "restore and maintain the chemical, physical, and biological integrity of the nation's waters." (None of the major implementation provisions of the act, however, were designed specifically to achieve this ambitious goal.) There follow two statements of operational objectives, noted above: the first calls for the elimination of all discharge of pollutants into navigable waters by 1985; the second, concerning the desired uses of the nation's water bodies, helps to define the nature of the benefits sought from controlling pollution and calls for "fishable and swimmable" waters by 1983. The act makes no provision for considering the costs of attaining either of these objectives.

The act also retains the system of state-established standards for in-stream water quality that was enacted in 1965. But under FWPCA-72 the states are now required to review these standards at least every three years and submit any revisions to the Environmental Protection Agency (EPA) for approval. The standards are to specify those uses of each water body that are to be protected and to establish maximum allowable concentrations of pollutants consistent with the designated uses. To be consistent with the objectives of the act, the minimum water quality standards must be sufficient to allow swimming and some types of fishing; but more strict standards are allowed—for example, standards that would protect pristine water quality in undeveloped rivers. Nothing in the act prohibits states from considering benefits and costs when they establish water quality standards.

In addition to these features, FWPCA-72 states that it is national policy "that the discharge of toxic pollutants in toxic amounts be prohibited," that federal financial assistance be provided for the construction of municipal treatment plants, and that areawide planning and technological innovation be encouraged.[10]

Means

The principal means for attaining water quality objectives under FWPCA-72 are the establishment and enforcement of technology-based effluent standards. These standards are quantitative limits imposed on all dischargers where the quantities are determined by reference to present technology. To put it simply, standards are set on the basis of what can be done with available technology, rather than what should be done to achieve ambient water quality standards, to balance benefits and costs, or to satisfy any other criteria. Since production processes, quantities and composition of waste loads, and treatment technologies vary substantially across industries, separate discharge standards must be developed for the different categories of industry. In FWPCA-72, these standards are referred to as effluent limitations.

Technology-based standards in general must spell out the degree of technological sophistication to be embodied in a standard. The limitations could be based on present standard operating practice, best current practice, best available or demonstrated technology, and so forth. If the legislation governing the standards called only for technological considerations, it would be a purely technology-based system. But often, as in the case of FWPCA-72, technological requirements are tempered somewhat by economic considerations and qualifications. These two aspects, the technological and the economic, illustrate a major difficulty in the technological approach to defining effluent standards. Although technology-based effluent standards may appear to be definitive and objective, in practice both the definition of the technology to be employed and the economic qualifications that are usually attached are imprecise and ambiguous. "Best practical," "best available," or "reasonable costs" are not objective, scientific terms. Such language grants tremendous discretion to the officials charged with the development of technology-based standards. In addition, the wording places a very large responsibility on those officials to interpret a bewildering assortment of engineering, scientific, and economic information.[11]

According to FWPCA-72, technology-based standards were to be achieved in two stages. By 1977, industrial dischargers were to be meeting effluent limitations based upon the best practicable control technology currently available (BPT). However, in determining what is practicable, the EPA was to consider "the total cost of application of technology in relation to the effluent reduction benefits to be achieved." Effluent limitations for publicly owned treatment works called for "secondary treatment" by 1977. By 1983, effluent limitations for industrial dischargers were to be based upon the "best available technology economically achievable" (BAT), while publicly owned treatment works were to meet effluent limitations based upon the "best practicable waste treatment technology."

The drafters of this legislation recognized the possibility that pollution loads in some rivers might be so severe that fishable and swimmable water quality could not be attained even though dischargers met the mandated effluent limitations. For that reason, the act called for the EPA to impose more stringent effluent limitations on point sources whenever this is found to be necessary to achieve the desired water quality standards. In other words, the act requires better than best in such cases.

The Clean Water Act of 1977 maintained the basic structure of technology-based standards and effluent limitations. But this act provided for somewhat different criteria than FWPCA-72 for establishing effluent limitations for conventional pollutants (organic material, suspended solids, bacteria, and pH) as well as for toxic pollutants. CWA-77 also modified some of the deadlines for achieving stated effluent limitations. After 1977, effluent limitations for conventional pollutants were to be established with reference to the "best conventional pollution control technology" (BCT) and were to be achieved by July 1, 1984. Dischargers of toxic pollutants were to meet best available technology effluent limitations either by 1984 or three years after the promulgation of such limits, provided that this was no later than 1987. The 1987 reauthorization act extended most of these deadlines once again, to 1989. Finally, the 1977 amendments permit the EPA to relax the secondary treatment requirement for municipalities that discharge into deep marine waters if they can show that fishable, swimmable, and drinkable water quality standards could still be attained.

Effluent limitations become the basis for discharge permits to be held by all dischargers. These permits limit the allowable discharges of individual polluters to the quantities that are consistent with the relevant technology-based effluent limitation. Permits were initially to be issued through the regional offices of the Environmental Protection Agency. If a state water pollution control agency satisfies certain conditions, however, it can take over responsibility for issuing permits and enforcing their terms. By November 1988, 39 states and territories had taken over responsibility for issuing permits.[12] Enforcement of permit terms is based on two kinds of compliance monitoring: the evaluation of self-reported discharge data, and on-site inspections by government enforcement personnel. (For further discussion of monitoring and enforcement problems, see chapter 7 in this volume.)

The Municipal Grant Program

Under federal law, the problem of pollution from municipal sewage treatment plants has been treated very differently from industrial pollution in one important respect. As an incentive to achieve compliance and to ease the financial burden on local government, under FWCPA-72 the federal government offered to pay up to 75 percent of the design and construction costs for municipal treatment plants. No such subsidy has ever been granted to private firms. The ability of municipalities to achieve secondary treatment is tied, therefore, to the level of funding for the municipal grant program. If grant funds are not available, lack of funds becomes a justification for failure to construct the necessary facilities.

In combination with matching funds from state and local governments, the $18 billion authorized for the period 1973 to 1975 would have been sufficient to fund the construction of $24 billion worth of treatment

facilities. Actually, through fiscal year 1976 less than 15 percent of the authorized funds had actually been spent, indicating substantial delays in planning projects, processing applications, and other procedures.[13]

Non-point Sources of Pollution

Most of the major provisions of the 1972 act deal with the control of discharges from point sources—that is, from factories and municipal sewage systems. Yet urban and nonurban non-point sources such as storm-water runoff, cropland erosion, and runoff from construction sites, pastures and feed lots, and woodlands are also major sources of pollution in many water bodies. Estimates of actual discharges of five pollutants in 1972, before the implementation of FWPCA-72, are shown in table 4–2. In aggregate, non-point sources in 1972 accounted for between 57 and 98 percent of total national discharges of these substances. Of course, to the extent that BPT, BAT, and BCT effluent limits have reduced point source discharges of these substances, the relative importance of non-point sources will have significantly increased since then.

Table 4–2. Estimates of National Discharges from Point and Non-point Sources, 1972, before FWPCA (millions of pounds per year)

	5-Day biochemical oxygen demand	Total suspended solids	Total dissolved solids	Total phosphorus	Total nitrogen
Point sources					
Industrial	8,252	50,355	290,184	353	559
Municipal	5,800	6,000	31,847	101	1,111
Total point sources	14,052	56,355	322,031	454	1,670
Non-point sources	18,901	3,422,321	1,536,458	2,986	12,480
National total	32,953	3,478,676	1,858,489	3,440	14,150
Non-point sources as % of total discharges	57%	98%	83%	87%	88%

Source: Leonard P. Gianessi and Henry M. Peskin, "Analysis of National Water Pollution Control Policies: 2. Agricultural Sediment Control," *Water Resources Research* vol. 17, no. 4 (August 1981) p. 804.

The principal tool for dealing with non-point sources in FWPCA-72 is section 208, which calls for the development and implementation of areawide waste treatment management plans. Section 208 makes explicit reference to non-point sources and also authorizes federal grants to share in the cost of developing areawide management plans. In 1977, the Clean Water Act Amendments authorized a program of grants, to be administered through the Soil Conservation Service of the Department of Agriculture, that would cover up to (but no more than) 50 percent of the costs to rural land owners of implementing and maintaining "best management practices" to control non-point source pollution. (Recent developments in the evolution of federal policy toward non-point source pollution are discussed in the section on economic issues, below.)

ACCOMPLISHMENTS OF THE PROGRAMS SINCE 1972

It is of great interest to know what has been accomplished as a result of all this legislative and regulatory activity and how well the machinery created by FWCPA-72 and subsequent amendments has worked in controlling pollution and improving water quality. For two reasons, this turns out to be a difficult question to answer with any degree of confidence.

The first reason has to do with the choice among the variety of measures of performance and accomplishment and with the availability of relevant data. For example, FWPCA-72 established a number of deadlines for the promulgation of effluent limitations, the issuance of permits, and other procedures. Thus one indicator of accomplishment would be how well the EPA has performed in meeting the deadlines. These are measures of agency performance. Another set of measures concerns the performance of polluters in complying with pollution control requirements. Have the required treatment systems been installed on time and according to the terms of permits? And have actual discharges been within the limits established by permits?

However, neither agency nor polluter-performance measures are directly related to the real objectives of the act—the reduction of pollution and the improvement of water quality. To evaluate these, it is necessary to examine what is known about changes in water quality. But since water quality can be measured by a number of parameters, all of which are variable in both space and time, any single aggregate measure is bound to be a simplification of a much more complex reality.

The second difficulty that arises in measuring accomplishments under federal law concerns distinguishing between improvements which have

been caused by implementation of the acts and improvements that have occurred for other reasons. The relevant question is not how much some measure of water quality has changed at some location between 1972 and 1985; this is a before-versus-after question. The right question is: How much better was actual water quality in 1985 than would have been observed in 1985 without the clean-up requirements imposed by the 1972 act, given the same economic conditions, weather, and rainfall? This is the with-versus-without question. To answer it, we would have to be able to predict what the discharges of pollutants would have been in the absence of the 1972 act, but with the provisions of the 1965 Water Pollution Control Act still in place.

Two kinds of forces are at work that would produce different answers to the with-versus-without and before-versus-after questions. On the one hand, economic growth and increases in industrial production would have likely led to increases in pollution discharges since 1972, in the absence of regulation. Thus, in the absence of the act, water quality might have actually declined between 1972 and 1985. If so, the answer to a before-versus-after question could lead to an underestimate of the accomplishments of the 1972 act. On the other hand, to the extent that state and federal agencies might have become more effective in utilizing the machinery created in 1965, reductions in pollution and improvements in water quality might have occurred after 1972 even in the absence of new legislation. If so, attributing all of the observed improvements (or prevention of degradation) to the 1972 act might overstate its true accomplishments. These problems should be kept in mind as we interpret the data presented below.

Agency Performance

FWPCA-72 required that all sources of water pollution reduce their discharges so as to be in compliance with effluent limitations by specific dates: BPT effluent limitations by 1977, BAT limitations by 1983. Before sources could comply with these requirements, however, the Environmental Protection Agency had two major tasks to perform. The first was to write the effluent limitations for a variety of industrial and municipal sources.[14] The second was to translate the terms of these effluent limitation guidelines into specific requirements to be imposed on each source, and to be spelled out in the permits or licenses issued to the sources. One measure of the accomplishments under the 1972 act, then, is how well the agency performed in writing the effluent limitation regulations and issuing permits to individual sources.

The 1972 act required that the EPA promulgate effluent limitations as regulations within one year of the date of enactment. This turned out to be an unreasonable requirement, and the EPA had not issued a single regulation by the deadline. In fact, the agency still had not promulgated all of the BPT regulations by 1977, even though that was the deadline for sources to be in compliance with the regulations; and there were more than 250 court cases challenging various provisions of those regulations which had been issued by then. It was at least in part in response to these problems that Congress in 1977 pushed back to July 1, 1984 the deadline for complying with BAT requirements for a specified list of toxic pollutants, and allowed for up to three additional years for compliance with BAT requirements for other toxic chemicals. As of 1988, the Environmental Protection Agency had issued BAT regulations for all but one category of discharger.[15]

In regard to the EPA's second task—translating effluent limitations into specific effluent reductions for individual sources and issuing permits to thousands of pollution dischargers—EPA reported that as of March 31, 1976 it had issued permits to only 67 percent of all industrial dischargers. The percentage was higher for major industrial sources—over 90 percent;[16] but many of these permits were written before the applicable effluent limitations were promulgated! Some of the issued permits were inconsistent with the effluent limitations eventually adopted. Permits are subject to periodic review and revision. But in the meantime some dischargers will be in nominal compliance with the terms of their permits but not in compliance with the applicable technology-based effluent standards. As of October 1982, the EPA had issued slightly more than 68,000 permits, more than 52,000 to industrial and other nonmunicipal sources and almost 16,000 to municipal treatment plants.[17] It has not been reported what percentage of these permits were consistent with the relevant technology-based effluent standards.

Dischargers' Performance

The second measure of accomplishment is the extent to which individual dischargers have complied with the terms of their relevant effluent limitations and discharge permits. It has been estimated that as of the 1977 deadline, about 80 percent of industrial dischargers were in compliance with the relevant BPT effluent limitations, and that by 1981, 96 percent of industrial sources would have been in compliance.[18] The compliance rate with the secondary treatment requirement from municipal dischargers was substantially lower in both 1977 and 1981. It has also been estimated that full compliance with the BPT regulations by industry

would result in, approximately, a 65 percent reduction in industrial discharges of oxygen-demanding organic material, an 80 percent reduction in industrial discharges of suspended solids, a 21 percent reduction in oil and grease discharges, and a 52 percent reduction in discharges of dissolved solids.[19] There are no comparable estimates of compliance rates or percentages of pollutant removed for the BCT and BAT effluent limitations.

The term "compliance" as used by the Environmental Protection Agency generally means the installation of treatment equipment capable of meeting the BPT effluent limitations when properly operated. These data on compliance do not say anything about actual discharges. In order to determine the degree of effective compliance, it is necessary to examine the discharges of polluters and to compare them with the terms of their permits and relevant effluent limitations.

The U.S. General Accounting Office (GAO) made an attempt to do this over an eighteen-month period in 1981–1982. However, because of limited resources for the study, the GAO had to rely on discharge data supplied by the dischargers rather than on independently determined or verified discharge data. Nevertheless, the GAO study indicates a significant noncompliance problem. Discharge data for about a third of all industrial and municipal dischargers in six states were examined. The results are summarized in table 4–3. Eighty-two percent of the sources

Table 4–3. Self-reported Noncompliance with Discharge Permits in Six States, October 1, 1989 to March 31, 1982

Type of discharger	Sample size	Sample as % of total number of sources	Noncompliance (% of sample)[a]	Significant noncompliance (% of sample)[b]
Municipal	274	34%	86%	32%
Industrial	257	36	79	16
Total	531	35	82	24

Note: The six states were Iowa, Louisiana, Missouri, New Jersey, New York, and Texas.

Source: General Accounting Office, *Waste Water Dischargers Are Not Complying with EPA Control Permits* (Washington, D.C., 1983) pp. 7–10.

[a] At least one reported monthly average reading exceeded the permit limit during the eighteen-month period.

[b] At least four consecutive monthly average readings exceeded the permit limit by 50 percent or more.

had at least one month of noncompliance during the eighteen-month period. Moreover, about 24 percent of the sample was in "significant noncompliance" for at least four consecutive months, during which dischargers exceeded permitted levels by at least 50 percent. The performance of municipal sources was poorer than that of industrial sources, especially in cases of significant noncompliance. The reasons offered for significant noncompliance included operation and maintenance deficiencies, equipment deficiencies, and treatment plant overloading.[20]

It must be emphasized again that the GAO findings were based on reports from the discharge sources themselves. Since such sources may have incentives to understate actual discharges, the extent of noncompliance could be even greater than indicated by the study.

Changes in Water Quality

It is important to try to determine whether the 1972 and 1977 laws have resulted in levels of water quality across the country that are better than they would have been, other things being equal, without these laws, and if so how much better. None of the available data can answer these questions conclusively; but we can draw some inferences from several sets of data. The data are of two types: observations of actual changes in water quality from monitoring stations and other sources, and predictions of changes in water quality made in response to changes in pollutant discharges that are based on water quality models which hold other things (such as the level of economic activity) constant.

It should be noted that water pollution control efforts have not been limited to the period after 1972, and that some states had significant regulatory and construction programs before 1972. These earlier efforts were apparently somewhat successful. The results of the EPA's first National Water Quality Inventory, conducted in 1973, indicated there had been significant improvements in most major waterways over the preceding decade, at least in regard to organic wastes and bacteria.[21] These improvements could not have been due to FWPCA-72.

Researchers at Resources for the Future (RFF) have modeled the effects of the 1972 amendments on several measures of water quality.[22] The RFF water quality network model is based on inventories of waste generated at point sources and estimates of actual removal rates as of 1972. The model also incorporates estimates of non-point source discharges. The inventories of wastes generated and discharged are combined with a model of how pollution moves through water to predict values for four water quality parameters at over 1,000 locations in the continental United States. Estimates of increased treatment levels as a consequence of BPT and secondary treatment effluent limitations can be used to predict changes in discharges, and hence the effects on measures of water quality across the country. The effects of other policies, such as non-point source control policies, can also be simulated with this model.

The results of two scenarios examined in the RFF study are summarized in table 4–4. The first scenario is based on the estimated 1972 discharges of polluting substances and predicts the percentage of locations achieving assumed water quality standards. As table 4–4 shows, 83 percent of all locations were predicted to have met the standard for dissolved oxygen, while 68 percent were predicted to have met the standard for biochemical oxygen demand. Since the principal effect of biochemical oxygen demand in water is to reduce dissolved oxygen concentrations, many locations having high biochemical oxygen demand must also have high assimilative capacity, so that the more critical dissolved oxygen standard is not violated. In this scenario relatively few locations were predicted to have attained the assumed standard for the nutrients phosphorus and nitrogen.

In the second scenario, the model predicts water quality at each location, assuming all point sources of pollution to be in compliance with

Table 4–4. Effect of BPT/ST Controls on Four Water Quality Parameters (1972 baseline)

	Percentage of locations meeting assumed water quality standard for:			
	Dissolved oxygen[a]	Biochemical oxygen demand[b]	Total phosphorus[c]	Total Kjeldahl nitrogen[d]
Estimated 1972 baseline discharges	83%	68%	27%	30%
Predicted, with BPT/ST controls	88	75	32	32
Percentage change in number of locations	6	10	19	7

Source: Gianessi and Peskin, "Analysis of National Water Pollution Control Policies," pp. 805–807.

[a] The standard is dissolved oxygen > 5.0 mg/1.

[b] The standard is 5-day biochemical oxygen demand ≤ 9.0 mg/l.

[c] The standard is total phosphorus ≤ .180 mg/l.

[d] The standard is total Kjeldahl nitrogen ≤ .90 mg/l.

the relevant BPT industrial effluent limitation or secondary treatment (ST) standard for municipal treatment plants. The model predicts that BPT/ST will result in increases in the number of locations meeting the standards for each of the four water quality parameters; but the absolute and percentage increases in locations meeting the standards are surprisingly small. The model predicts only a 6 percent increase in the number of locations satisfying the dissolved oxygen standard, but this is in large part because of the high percentage of locations already meeting the standard before application of BPT/ST. (As table 4–2 showed, uncontrolled non-point sources make a significant contribution to the total load of biochemical oxygen demand.) On the other hand, for those two parameters where there is greatest room for improvement—phosphorus and nitrogen—the application of BPT/ST has a relatively small effect on the number of locations in violation. This is because the point sources affected by BPT/ST are relatively unimportant sources of these pollutants (see table 4–2). In summary, to the extent that the RFF model accurately predicts water quality, it appears that the first, or BPT/ST, phase of FWPCA-72 has had relatively little effect on these measures of water quality in many areas.

Measures of water quality such as dissolved oxygen or total phosphorus may not have much meaning to most people. What matters most to them is how changes in these measures affect various uses of a water body. One such use of rivers and lakes is recreational fishing. To the extent that reduced pollution and improved water quality result in more recreational opportunities and higher-quality recreation, recreational fishermen are made better off. Recently a method was developed for classifying water bodies by the quality of fishing opportunities they present, and for translating changes in measures of water quality, as predicted by the RFF water quality network model, into changes in the number of acres of surface waters available for various categories of fishing.[23] According to this approach, a body of water can be placed in one of four categories, depending upon water temperature, pH, and the concentration of dissolved oxygen and total suspended solids. The four categories are: (1) not fishable; (2) fishable for warm-water rough fish, such as carp and catfish; (3) fishable for warm-water game fish/panfish, such as perch and bass; (4) fishable for cold-water game fish, such as trout.

Using the estimates of actual discharges in 1972, this model predicted that only 4.2 percent of the waters included in the model fell into the unfishable category in 1972. The prediction was consistent with the results of an RFF survey of state recreation and fishery officials which showed that only 7.4 percent of the nation's waters were judged to be unfishable as of 1972.[24]

Estimates of the total U.S. fishable water area in 1972 and of the allocation of the total fishable area to the three categories of fishing activity (with and without pollution controls) are shown in table 4–5. The implementation of BPT and secondary treatment is predicted to increase the total fishable area by only 0.35 percent. The major benefit of the BPT/ST regulations comes from improving the quality of fishing in already fishable areas. The area of water suitable only for warm-water rough fish is predicted to decline by 66.8 percent, while the areas of warm-water game fish and cold-water game fish increase by 8.0 percent and 11.7 percent, respectively.

These results from modeling exercises are consistent with actual observations of water quality and analyses of water quality monitoring data. For example, the Association of State and Interstate Water Pollution Control Administrators asked each state's water pollution control agency to estimate whether the quality of the state's streams, lakes, and estuaries was adequate for designated uses as of 1982, and whether water quality had changed over the preceding decade. While responses were based only partly on systematic monitoring of in-stream water quality, the overall results point to relatively good water quality in 1982 and relatively little improvement since 1972.[25] These results are summarized in table 4–6. As shown there, 64 percent of stream miles and 84 percent of lakes and reservoirs had water quality sufficient to fully support the designated

Table 4–5. Estimates of Changes in Fishable Waters Due to BPT/ST Pollution Control (in thousands of acres)

	Total fishable acres	Warm-water rough fish acres	Warm-water game fish acres	Cold-water game fish acres
Estimated 1972 baseline discharges	30,615	3,429	20,941	6,245
Predicted, with BPT/ST controls	30,721	1,137	22,611	6,974
Percentage change in acres	+.35%	−66.8%	+8.0%	−11.7%

Source: William J. Vaughan and Clifford S. Russell, *Freshwater Recreational Fishing: The National Benefits of Water Pollution Control* (Washington, D.C., Resources for the Future, 1982) table 2–9, p. 54.

Table 4–6. Levels and Trends of Water Quality as Reported by State Water Pollution Control Agencies

I. Water quality levels, 1982

	Fully support designated uses (%)	Partially support designated uses (%)	Do not support designated uses (%)	Quality not known (%)
Streams (miles)	64	22	5	9
Lakes and reservoirs (acres)	84	10	3	3

II. Water quality trends, 1972–1982

	Improving (%)	Unchanged (%)	Declining (%)	Not known (%)
Streams (miles)	11	67	2	20
Lakes and reservoirs (acres)	2	62	10	26
Estuaries (square miles)	22	74	3	1

Source: Based on The Conservation Foundation, *State of the Environment: An Assessment at Mid-Decade* (Washington, D.C., The Conservation Foundation, 1984) pp. 106–108.

uses—that is, drinking water supply, fishing, or contact recreation. Only 5 percent of stream miles and 3 percent of lakes and reservoirs were so polluted that they could not support designated uses. The majority of all water bodies were reported as having unchanged water quality over the decade. Trends of improving water quality were most common for estuaries (22 percent) and least common for lakes and reservoirs (2 percent); and 10 percent of the lakes and reservoirs were reported to have trends of declining water quality.

The only nationwide system for gathering data on various measures of water quality is the National Stream Quality Accounting Network operated by the U.S. Geological Survey (USGS). Statistical analyses of USGS data show similar trends in water quality; some of these results are shown in table 4–7. Overall, the percentage of readings in violation of a dissolved oxygen standard of 5.0 mg/liter is low, varying from 4 percent to 11 percent per year over the period 1975–1980. Only 11.2 percent of all

Table 4–7. Trends in Measured Water Quality

Water quality measure	Percentage of readings in violation of standard[a]						Percentage of stations showing trend for 1974–1981	
	1975	1976	1977	1978	1979	1980	Improvement	Decline
Fecal coliform bacteria	36	32	34	35	34	31	7.6	4.3
Dissolved oxygen	5	6	11	5	4	5	11.2	6.1

Note: This table is based on data from the U.S. Geological Survey's National Stream Quality Accounting Network (NASQAN).

Sources: Council on Environmental Quality, *Environmental Quality—1981* (Washington, D.C., 1981) table A-55, p. 24; Council on Environmental Quality, *Environmental Quality—1982* (Washington, D.C., 1982) pp. 39–41.

[a] For fecal coliform bacteria, a violation is above 200 cells per 100 ml.; for dissolved oxygen, a violation is less than 5.0 mg/l.

monitoring stations with sufficient data show a statistically significant trend of improving water quality as measured by dissolved oxygen, while 6.1 percent show a declining trend. About a third of all readings of fecal coliform bacteria have been in violation of water quality standards, and there has been little change in this percentage over the interval 1975–1980. As in the case of dissolved oxygen, stations with improving trends in bacteria counts outnumbered those with trends of declining water quality. But it is most striking that almost 90 percent of all stations report no trend for the years 1974–1981.

A recently published comprehensive analysis of trends in a large number of water quality measures provides similar results for the period 1974–1981.[26] Stations showing improvements in bacteria and dissolved oxygen levels outnumbered stations showing deterioration (substantially, in the case of bacteria); but fewer than 20 percent of the stations showed improvements in these measures of water quality. As for phosphorus and suspended sediments, the percentage of stations showing improved levels (11 percent and 14 percent, respectively) were approximately equal to the percentages showing deterioration (13 percent and 13 percent). Stations showing deteriorating levels of nitrates outnumbered those showing improvements by 4½ to 1. The authors attribute the latter results largely to increases in fertilized agricultural acreage and to atmospheric deposition of nitrates in eastern watersheds. The other significant trends reported were increases in salinity (due to human wastes and the use of salt on highways) and toxic trace elements (especially arsenic and cadmium, due to atmospheric deposition), and a decrease in lead.

Summary of Accomplishments since 1972

This review of the available evidence on accomplishments since 1972 suggests the following conclusions:

1. As measured by ability to support designated uses or to meet certain water quality standards, average water quality was not so bad in the United States in 1972. Large areas of water could support designated uses or meet water quality standards. However, some localities—particularly large, industrialized metropolitan areas—were experiencing serious water pollution problems.
2. There has been some improvement in water quality since 1972. In terms of aggregate measures or national averages, it has not been dramatic. But there are local success stories of substantial cleanup in what had been seriously polluted water bodies.
3. Despite passage of the Federal Water Pollution Control Act of 1972, we are losing ground to pollution in some areas. Some water bodies show trends of declining water quality.
4. It is possible that the lack of dramatic improvement in water quality has been due in part to slow implementation of the major features of FWPCA-72. Even the most recent data, for 1982, do not reflect the potential contributions of the BAT and BCT effluent limitations. And given the slow rate of promulgation of BCT regulations and low rates of at least initial compliance, especially by municipalities, we would not expect to see trends of improving water quality much before, say, 1978.

Two qualifications to these conclusions must be kept in mind. The first is that when we have looked at trends in water quality and the ability of waters to meet designated uses, we have been answering a before-versus-after question. We cannot tell from this evidence whether pollution might have gotten much worse without FWPCA-72 as a result of economic growth and increases in the production and discharge of polluting materials.

Second, all of the above conclusions regarding water quality refer only to the conventional pollutants and related measures of water quality, such as bacteria, nutrients, and dissolved oxygen. The picture could be different if it included toxic pollutants, but there are virtually no data from which trends in toxic pollution could be established. This deficiency must be remedied.

ECONOMIC ISSUES AND PROBLEMS

In examining federal water pollution control policy from an economic perspective, we look at the costs of controlling water pollution as well as what we are getting for our money—that is, the benefits. Pollution control is costly. Devoting more of society's scarce resources of labor, capital, and administrative and technical skills to pollution control necessarily means that less is available to do other things also valued by society—either collectively or as individuals. Because pollution control is costly, it is in society's interest to be "economical" in its selection of pollution control objectives.

There are two senses in which this is true. First, whatever pollution control objectives are chosen, the means of achieving them should be selected so as to minimize cost. Using more resources than are necessary to achieve pollution control objectives is wasteful; yet as we will show, current water pollution control policies are wasteful in several aspects. Second, society should be economical in its choices of environmental objectives. If we are to make the most of our endowment of scarce resources, we should compare what we receive from devoting resources to pollution control with what we give up by taking resources from other uses. We should undertake pollution control activities only if the results are worth more in some sense than the values we forgo by diverting resources from other uses. This is the universal core of truth in the benefit-cost approach to pollution control policy.

Government intervention to control pollution is justified on grounds of economic efficiency if the beneficial effects (broadly defined) to society as a whole from such action outweigh the costs. One can question whether this justification should be accepted as a presumption for any and all proposed environmental targets (such as swimmable and fishable water, and elimination of discharges), or whether evidence should be gathered and judgments rendered on a case-by-case basis. The economic approach to pollution control policy proceeds on the premise that not all interventions in behalf of environmental protection are desirable per se. Some may cost more than they are worth—not only in terms of the private calculus of market values, but also in terms of individual and social welfare. Examination of costs and beneficial effects should become an integral part of the process of establishing pollution control objectives.

This section begins by reviewing available information on the benefits and costs of federal water pollution control policy. We then search for ways in which the present set of policies might be modified to improve the relationship between benefits and costs. This takes us into a consideration of modifications of pollution control objectives and requirements where costs at the margin exceed benefits, and into a search for ways to reduce the costs of achieving given pollution control targets—that is, improved cost-effectiveness. We then apply some of the principles and insights developed here to two specific aspects of current policy, the municipal construction grant program and the program for controlling non-point sources of pollution.

A Comparison of Benefits and Costs

In order to provide a conceptually sound estimate of the benefits of a pollution control regulation, it is necessary to have information on four separate relationships (as pointed out in chapter 2). The first is that between the specific details of the regulation and changes in the quantities of polluting substances discharged. The second is the relationship between changes in the quantities discharged and changes in the relevant measures of water quality. The third is that between changes in water quality and changes in the uses people make of that water (for example, an increase in recreational activity or a reduction in treatment costs for municipal and industrial wastes might result in changed water quality). The fourth relationship is that between changes in uses and the monetary values which people place on these changes. Monetary values can be determined by applying a variety of economic models to estimate the maximum sums of money individuals would be willing to pay in order to attain the improved water quality and the increased uses which would be made possible.[27] For a set of pollution control regulations covering all industrial and municipal sources, it is necessary, in principle, to analyze each of these sets of relationships for each type of discharger and polluting substance, and to estimate the resulting water quality changes and changes in use and value for each of the affected water bodies. The national benefits of the regulatory policy can then be estimated by aggregating over all affected water bodies.

Unfortunately, there have been no studies of the aggregate national benefits of FWPCA-72 that deal in a fully satisfactory manner with all four sets of relationships. Some studies deal carefully with one or two of the relationships and/or one particular type of use, or examine the benefits of controls at one particular water body. Lacking fully satisfactory national aggregate benefit estimates, the analyst who wishes to make a benefit-cost comparison for the overall federal water pollution control policy must do so through some kind of synthesis and extrapolation from the most soundly based of existing studies.

One such estimate of national benefits was prepared for the Council on Environmental Quality (CEQ) in 1979 and revised in 1982.[28] The revised estimate is discussed here. This estimate was based on a review of approximately twenty empirical studies of various categories of damages resulting from water pollution and on estimates of benefits of a number of water pollution control policies. In the CEQ study, estimates of benefits were provided for four broad categories: recreation, nonuser benefits stemming from aesthetic and ecological changes, improved productivity of commercial fisheries, and diversionary uses including municipal and industrial water supplies. These estimates are presented in table 4–8. There is a substantial range of uncertainty in the estimates because of

Table 4–8. Benefits in 1985 of Achieving BPT and BAT/BCT Pollution Control (in billions of 1984 dollars)

Category	Range	Most likely point estimate
Recreation		
Freshwater fishing	$ .7– 2.1	$ 1.5
Marine sports fishing	.1– 4.5	1.5
Boating	1.5– 3.0	2.2
Swimming	.3– 3.0	1.5
Waterfowl hunting	.0– 0.5	.2
Subtotal	2.6–13.1	6.9
Nonuser benefits		
Aesthetics, ecology, and property value	.7– 5.9	1.8
Commercial fisheries	.6– 1.8	1.2
Diversionary uses		
Drinking water/health	.0– 3.0	1.5
Municipal treatment costs	.9– 1.8	1.3
Households	.2– 0.7	.4
Industrial supplies	.7– 1.4	.9
Subtotal	1.8– 6.9	4.1
Total	$5.7–27.7	$14.0

Note: 1984 dollars have been adjusted using the implicit price deflator for the gross national product.

Source: A. Myrick Freeman III, *Air and Water Pollution Control: A Benefit-Cost Assessment* (New York, Wiley, 1982) table 8-3, p. 161, and table 9-1, p. 170.

limitations in the available data and knowledge, and in some cases differences in the results of studies analyzing similar phenomena.

As indicated in table 4–8, the national benefits to the U.S. population in 1985 were estimated to be at least $5.7 billion per year (in 1984 dollars), although they could have been as high as $27.7 billion per year. The most likely value as judged by the analyst was $14.0 billion per year. Of this total, about half was due to improvements in water-based recreation opportunities. It would have been useful for the purposes of policy analysis to have separate estimates of the benefits of BPT (the first stage of controls called for in FWPCA-72) and of the incremental benefits associated with the subsequent BAT/BCT requirements. However, given the nature of the studies on which this synthesis was based, there was no reasonable basis for allocating the total to the two separate policy steps.

Since the true costs of meeting the BPT and BAT/BCT standards depend on a myriad of technical and financial decisions by individual dischargers, and since the costs involve resource outlays made over a substantial period of time, no single figure can precisely and accurately measure the cost of federal water pollution control policy. Two sets of estimates of costs can provide a basis for establishing a range of true costs for comparison with the benefit figures discussed above. In both cases, the caveat about frequent confusion over the relationship between expenditures and costs should be remembered.

The first set of estimates was prepared by the Council on Environmental Quality on the basis of the EPA's estimates of the engineering costs of meeting the 1972 requirements. This study did not take into account the 1977 amendments to FWPCA. For the years 1978 and 1988, the study presented separate estimates of operation and maintenance costs and the annualized capital costs of in-place equipment (interest and depreciation), as well as a total annual cost. This study also provided an estimate of cumulative annual costs for the decade 1979–1988.[29] These cost data are summarized in table 4–9. The public costs indicated are those associated with the collection and treatment of municipal wastes. They constitute an increasing share of the total water pollution control costs over the decade; for the decade as a whole, public costs amount to about half of the total cost. Annual costs are projected to almost double, rising from $17.4 billion per year in 1979 to $33.4 billion per year by 1988. The average annual cost over the decade (without discounting) is $23.2 billion per year.

The Environmental Protection Agency has recently published its own more up-to-date set of estimates of the annual cost of complying with federal water pollution control laws.[30] The EPA's estimates of total annualized costs for 1981 and the annual average of these costs over the

Table 4–9. CEQ Estimate of Incremental Pollution Abatement Costs for 1979–1988 (in billions of 1984 dollars)

	Operation and maintenance	Capital costs[a]	Total costs
	I. 1979		
Public	2.3	5.9	8.2
Private	5.1	4.1	9.2
Total	7.4	10.0	17.4
	II. 1988		
Public	4.5	13.7	18.2
Private	7.8	7.4	15.2
Total	12.3	21.1	33.4
	III. Average annual cost, 1979–1988		
Public	3.4	8.1	11.5
Private	6.1	5.5	11.7
Total	9.5	13.6	23.2

Note: Incremental costs are those made in response to FWPCA-72. 1984 dollars have been adjusted using the implicit price deflator for the gross national product.

Source: Council on Environmental Quality, *Environmental Quality—1980* (Washington, D.C., 1980).

[a] Interest on accumulated investment and depreciation.

decade 1981–1990 are given in table 4–10. These estimates are based on the regulations in place as of December 1982, thus any costs to be incurred in meeting BAT regulations promulgated after that date are not included. The EPA cost estimate is somewhat higher than that of the Council on Environmental Quality, a difference that seems reasonable since the former includes the costs of at least some of the BAT requirements imposed by the 1977 amendments. The major difference between the two sets is EPA's much higher estimate of the public-sector costs of compliance with the federal law. On the other hand, EPA's estimates of private-sector costs are somewhat below those of the Council on Environmental Quality.

On the basis of these two sets of estimates, it is reasonable to infer that the annual estimated costs of federal water pollution control policy for the year 1985 lie somewhere in the neighborhood of $25 to $30 billion (in 1984

Table 4–10. EPA Estimate of Pollution Control Costs for 1981–1990 (in billions of 1984 dollars)

	1981	Average (per-year) cumulative cost, 1981–1990
Government	$17.7	$21.9
Private	7.5	8.9
Total	25.2	30.8

Note: 1984 dollars have been adjusted using the implicit price deflator for the gross national product.

Source: Environmental Protection Agency, *Final Report: The Cost of Clean Air and Water*, Report to Congress, 1984 (Washington, D.C., 1984) table 3, pp. 15–16.

dollars). This is substantially higher than the most likely estimate of the benefits to be realized in 1985. In fact, the range of the estimates for benefits ($5.7–27.7 billion) barely overlaps the bottom end of the estimated range for costs. Thus it seems unlikely that benefits would be found to exceed the costs if the true values of both could be ascertained. On balance, therefore, it appears that the benefit-cost relationship for the present water pollution program is unfavorable.

However, several qualifications must be attached to this conclusion. The first—a potentially important one—is that these estimates of benefits do not include the benefits to human health, recreation, or ecosystems that might be realized by the control of toxic substances and metals in effluents. Yet the EPA's estimate of the costs of control in 1985 include those for controlling discharges of these substances. To the extent that these substances now cause significant damages, the benefit-cost relationship of water pollution policy would be improved by including the benefits of such control in the estimates.

Another qualification is that it is generally believed that cost estimates based on engineering design studies are biased upward, for at least two reasons.[31] First, these estimates are based on the assumption that there would be no increases in the prices of products or decreases in output levels in response to higher pollution control costs. Yet the prices of those goods having the highest pollution control costs are likely to rise relative to the general price level, thus inducing consumers to reduce purchases of the goods. In the long run, then, as the output of goods with high pollution control costs decreases, the total cost of controlling the pollution associated with those goods will also decrease. Second, engineering estimates are based on average or typical plant conditions. But the actual pollution control techniques adopted by individual firms will depend on their unique circumstances, to the extent that EPA regulations allow some choice. If firms can adapt the available engineering techniques to their own particular circumstances or find other techniques for meeting the standards, their selection of lower-cost techniques will result in a lower aggregate cost to pollution control. And third, engineering estimates assume there will be no technological change or innovation in the pollution control industry. Yet the high costs of pollution control themselves provide a significant economic incentive for technological change and innovation which has the effect of lowering aggregate pollution control costs. To the extent that true pollution control costs turn out to be lower than the estimates given above, the benefit-cost relationship of present policies will turn out to be more favorable than shown here.

Nevertheless, even if the true benefits turn out to be 20 percent higher and the true costs 20 percent lower than the estimates above, it seems likely that costs would still outweigh benefits. For this reason it is important to seek ways to modify present policies so as to improve the benefit-cost relationship. Broadly speaking, there are two such avenues to be investigated. The first involves lowering targets or pollution control requirements where, at the margin, the costs of current controls substantially exceed the benefits. The second involves seeking ways of reducing the costs of achieving the existing goals. We explore these avenues in the next two subsections.

Adjusting the Targets and Standards

Present water pollution control policy establishes a national interim target of fishable and swimmable water quality and an ultimate stated objective of zero pollutant discharge. It also imposes pollution reduction requirements on individual dischargers based on effluent guidelines that are uniform within industrial categories. If we were to adopt as an alternative the principle that pollution control policies should be designed to maximize the net benefits from pollution control activities, then pollution control requirements for individual dischargers would emerge as the result of a two-part analytic process. The first part would involve the establishment of a set of water quality standards for each water body and segment of river, such that in each case the incremental or marginal benefits of raising water quality to that point would just equal the marginal cost of doing so. In those cases where marginal pollution control costs were high, the resulting water quality standard might be lower than the fishable-swimmable national target. But in other cases this economic benefit-cost approach might lead to very high standards for water quality.

The second part of the analytic process would involve determining the individual effluent reductions necessary to meet the water quality standards for each water body and each segment of river. These reductions might vary across dischargers not only because of differences in industrial processes and control technologies, but also because of differences in costs and impacts on water quality. This approach to policymaking could save resources by imposing less stringent effluent limitations where the marginal costs of achieving fishable-swimmable water quality were greater than the marginal benefits of doing so. It should be noted that economically based water quality standards would have to be reviewed periodically and revised to reflect changes in the economic and other factors determining benefits and costs.

The Water Quality Act of 1965 provided a framework through which such an economic approach could have been implemented. But Congress chose to replace that framework with the present one in 1972. At that time, Congress established the National Commission on Water Quality to consider, among other things, whether the fishable-swimmable water quality target should be retained. The commission recommended that the target be retained, and Congress implicitly accepted this recommendation in 1977 when the statement of goals and the system of technology-based effluent standards were left intact. The choice of a criterion for policymaking is essentially a political question. The adoption of an explicitly economic framework for policymaking would be a major change in philosophy. At least for now, Congress has settled the question in favor of the present technology-based approach.

The National Commission on Water Quality went on to recommend in its 1976 report that more stringent effluent limitations be applied selectively where they can be shown to have a significant beneficial effect in achieving water quality standards. The commission also called for further study to consider whether uniform application of BAT or other more stringent effluent limitations was justified and desirable. In discussing these recommendations, the commission stated that "wise use of total national resources would support the imposition of only those levels of control or treatment that actually produce the intended result. Additional investment with marginal identifiable benefits or improvements could operate to the detriment of competing demands of other worthwhile national programs, such as energy conservation and air pollution control."[32]

Even if the present framework for policymaking is left essentially unchanged, there remain several opportunities for making the kinds of adjustments suggested by the commission. One possibility is to authorize more formal and substantive analysis of economic factors in establishing BAT and BCT effluent limitations. Congress has given some latitude to the EPA in establishing BPT and BCT effluent limitations. For example, in describing the factors to be taken into account in establishing BCT standards, Congress called for "consideration of the reasonableness of the relationship between the costs of obtaining the reduction in effluents and the effluent reduction benefits derived. . . ."[33] This language would appear to allow the balancing of benefits and costs in setting standards. However, the language defining BAT requirements includes a reference to the cost of achieving effluent reductions but does not explicitly authorize a consideration of benefits.

A second way in which economic considerations could influence pollution control standards would be to allow officials who write individual discharge permits greater latitude to deviate from the effluent limitation guidelines after case-by-case comparisons of water quality benefits and costs. For example, where fishable-swimmable water quality has already been achieved, little may be gained by imposing additional BCT or BAT requirements on sources discharging into that water body. Yet present law does not provide any authority for waiving BAT requirements where fishable-swimmable water quality targets are already being met.[34]

The 1977 amendments to FWPCA-72 did create one opportunity for relaxing pollution requirements on publicly owned treatment works. One section of the 1977 act permits a waiver of secondary treatment requirements for municipal facilities that discharge into ocean waters where fishable-swimmable requirements are satisfied and where it can be shown that the assimilative capacity of the ocean waters will allow for the absorption and dispersion of these wastes without impairing other uses of the water.

There are also likely to be bodies of water where BAT and BCT limitations will not be sufficient to achieve fishable-swimmable waters. This would mean that the costs of BAT and BCT would be incurred without realizing the benefits of fishable-swimmable water. One reason for such nonattainment is that even with BAT and BCT in place, the total volume of industrial and municipal discharges might still be too great to permit achievement of the fishable-swimmable standard. The 1972 act actually anticipates a situation of this kind by requiring even stricter effluent limitations for these cases. However, the EPA administrator is required to examine the relationship between costs and benefits and may choose not to require the additional effluent limitation if he or she finds that there is no reasonable relationship between economic and social costs and the benefits to be obtained.

A more sophisticated approach to dealing with these cases would be to take advantage of seasonal variations in designated water uses or seasonal variations in water quality. For example, where cold weather precludes water-based recreation, water quality standards and pollution control requirements might be relaxed during winter to save on pollution control costs. Or where existing water quality standards are violated only during some seasons (as when there is low water flow or high water temperature), the most stringent control requirements could be imposed only during the problem seasons.[35] It must be acknowledged, however, that such intermittent controls may be more difficult to administer.

A second possible reason for not attaining fishable-swimmable water quality with BAT and BCT limitations could be pollution from non-point sources, since these are not presently covered by any effluent limitations, technology-based or otherwise. The larger the non-point sources of pollution relative to point sources in any water body, the less impact control of point sources will have on ambient water quality. Thus significant control of non-point sources may be necessary to achieve fishable-swimmable water in these cases. And it may be possible to substantially reduce the costs of achieving fishable-swimmable water by imposing relatively more strict requirements on non-point sources while relaxing the pollution control requirements on point sources.

Improving Cost-Effectiveness

By cost-effectiveness economists mean the degree to which an activity, such as controlling pollution, is carried out so as to achieve the stated targets—say, water quality standards—at the lowest possible total cost. The importance of achieving cost-effective pollution control policies should be self-evident. Any cost savings that can be achieved frees resources which can be used to produce other goods and services of value to people. If some change in the allocation of clean-up requirements among dischargers results in a lower total cost of controlling pollution without degrading water quality, then society is clearly better off.

A pollution control policy is cost-effective only if it allocates the responsibility for cleanup among sources so that the incremental or marginal cost of achieving a one-unit improvement in water quality at any location is the same for all sources. Differences in the marginal cost of improving water quality can arise both from variations in the marginal cost of treatment or waste reduction across sources and from variations among sources in the effects of lower discharges on water quality.

To illustrate the first point, suppose that two adjacent factories discharge the same substance into a lake. This means that an *x*-unit decrease in discharges gives the same incremental benefit to water quality whether it is achieved by factory A or factory B. Now suppose that in order to achieve fishable-swimmable water quality in the lake, discharges of the polluting substance must be reduced by 50 tons per day. One way to achieve the target is to require each factory to clean up 25 tons per day. But with this allocation of clean-up responsibility, what if factory A's marginal cost of cleanup is $10 per ton per day while at factory B the marginal cost is only $5 per ton per day? Allowing factory A to reduce its cleanup by one ton per day saves it $10. If factory B is required to clean up an extra ton, total cleanup is the same and the water quality standard is met, and the total cost of pollution control is reduced by $5 per day. Additional savings are possible by continuing to shift clean-up responsibility to B and away from A (thus reducing A's marginal cost). This should continue until B's rising marginal cost of control is made equal to A's now lower marginal cost. When discharges of a substance have different impacts on water quality depending upon the location of the source, this circumstance must also be taken into account in finding the least-cost or cost-minimizing pattern of discharge reductions.

From an economic standpoint, a major criticism of technology-based standards is that they are virtually certain to result in a higher than necessary total cost for any particular level of water quality. There is nothing in the logic or the procedures for setting technology-based limits to assure that the conditions for cost minimization will be satisfied. Since the marginal cost of control is not systematically taken into account, technology-based standards are not likely to result in equal marginal costs of reducing discharges across different sources of the same substance. One analysis of the marginal cost of removing oxygen-demanding organic material with the BPT standards found a thirtyfold range of marginal costs within the six industries examined.[36] In this instance spending an extra dollar for treatment in a low-cost industry would buy thirty times more pollution removal than spending the same dollar in a high marginal-cost industry. One set of studies prepared for the National Commission on Water Quality showed potential for substantial cost savings (30 to 35 percent for the nation as whole) through the selection of cost-minimizing effluent reductions to achieve the same water quality improvement.[37]

Several possible modifications of present pollution control policy would go a long way toward improving its cost-effectiveness. In the past, most economists have favored placing a tax or charge on each unit of each pollutant discharged and allowing each discharger to choose that degree of cleanup which minimizes its total cost (that is, clean-up cost plus tax).[38] The effluent charge strategy has long been attractive to economists because it provides a certain and graduated incentive to firms by making pollution itself a cost of production. And it provides an incentive for innovation and technological change in pollution control. A system of effluent charges can also make a major contribution to cost-effectiveness. Where several sources are discharging into the same water body, effluent charges would induce them to minimize the total cost of achieving any given reduction in pollution, because each discharger would control discharges up to the point where its marginal or incremental cost of control is equated with the given charge. If all dischargers face the same charge, they will have equated their marginal cost of pollution control. This is the condition for cost-minimization in reducing charges. There is no reallocation of responsibilities for reducing discharges that would achieve the same total reduction at a lower total cost.

It is also possible to use water quality models to design systems of charges that are differentiated by location in order to take into account the differing effects of discharges on water quality across sources. With the appropriate set of differentiated charges, the total cost of meeting any set of water quality standards within a river basin or region can be minimized. Some studies have shown that costs might be reduced by as much as 20 to 50 percent by appropriately designed systems of effluent charges.[39] To calculate systems of optimal differentiated effluent charges for a particular river requires lots of data and sophisticated economic and water quality modeling. It is probably not practical to recommend that such calculations be carried out for all sources of pollution on all rivers and streams in the United States. But a simpler effluent charge system, such as the one adopted in the Federal Republic of Germany in 1976, may be feasible and desirable.[40]

Another approach to water pollution control having essentially the same incentive and cost-minimizing effects is a system of tradable or marketable discharge permits. Under such a system, the pollution control authority would issue a limited number of pollution permits or tickets. Each ticket would entitle its owner to discharge one unit of pollution during a specified time period. The authority could either distribute the tickets free of charge to polluters or other parties on some basis, or auction them off to the highest bidders. Dischargers might also buy and sell permits among themselves. As suggested in chapter 2, a system of marketable discharge permits could result in an allocation of pollution rights that would meet the ambient water quality standards while minimizing the total costs of pollution control.

When dischargers are located at different points along a river, the terms at which permits are exchanged would have to reflect the different locations of sources and the varying impacts of their discharges on water quality. One study of a proposed system of transferable discharge permits for industries and municipalities along the Fox River in Wisconsin showed the potential for cost savings of 29 to 66 percent when compared with the uniform application of best practicable control technology and secondary treatment. The magnitude of the cost savings depended on water temperature and flow.[41]

One advantage of a system of transferable discharge permits in comparison with effluent charges is that the former represents a less radical departure from the existing system. Since all sources are presently required to obtain permits specifying maximum allowable discharges, it would be a relatively straightforward matter to rewrite permits in a divisible format (ten one-ton permits instead of one ten-ton permit) and allow exchanges to be made. A more modest step would be to allow two (or more) sources to propose a reallocation of clean-up requirements between them if they found it to their mutual advantage and if there were no degradation of water quality. A source having low incremental costs of control should be willing to increase its cleanup provided another firm compensated it for its increase in pollution control costs. And a source having high pollution control costs would find it cheaper to pay another source rather than clean up itself. The permit-writing authority should be willing to rewrite the permits for such sources if it found that water quality would not be lowered.

A first step toward obtaining the economic advantages of transferable discharge permits is the application of the "bubble" concept to water pollution control.[42] In a major industrial facility such as an integrated steel mill there may be several separate activities or processes going on, each subject to a different BCT or BAT requirement. Many of these processes discharge the same substances, yet the incremental costs of pollution control may be quite different across activities.[43] As a result, the total cost of controlling the aggregate discharge from the plant is often higher than necessary. In such cases, plant managers should be allowed to adjust treatment levels at different activities if they can lower total treatment costs, as long as the total amount of a pollutant discharged from the plant does not exceed the aggregate of the effluent limitations for individual processes. The Environmental Protection Agency is now allowing such bubble tradeoffs at integrated steel mills, provided that the tradeoffs result in a net reduction of the total amount of pollutants discharged.[44] Perhaps present law should be modified to facilitate similar intraplant trades in all industrial categories.

In some river basins, non-point sources are major contributors of at least some polluting substances. There, control of non-point sources will be required if fishable and swimmable water is to be achieved. But even where control of non-point sources is not necessary, such control might still be desirable on grounds of cost-effectiveness if the marginal cost of initial controls on non-point sources is lower than the marginal cost of additional controls (BAT or BCT) on point sources. We will say more about cost-effectiveness in the section on non-point sources of pollution below.

The Municipal Construction Grant Program

A major issue before Congress in early 1987 when it considered reauthorization of the Clean Water Act was what to do with the municipal construction grant program. In 1981 Congress had authorized federal cost-sharing grants of $2.4 billion per year through fiscal year 1985, and key congressional leaders believed they had an agreement with the administration to continue funding at this level through 1991.[45] The Reagan administration, acting through the Office of Management and Budget, proposed the elimination of the federal grant program by 1990. The administration proposal would have resulted in a total grant authorization over the period 1985–1989 of $8.4 billion; given a federal cost-sharing rate of 55 percent, this would have enabled total federally supported spending of $15.3 billion. The reauthorization bills approved by the two houses of Congress in 1985, although different in some details, provided for federal contributions two to two-and-a-half times greater than those in the administration proposal. The congressionally proposed contributions were to be a combination of grants and federal support of state-managed revolving loan funds. Funding levels of both the administration and congressional proposals are shown in table 4–11. It was the Senate's plan that was enacted in the Water Quality Act of 1987.

The Environmental Protection Agency recently estimated that capital spending of $40.6 billion would be required to meet the full needs of municipal sewage systems for correcting problems of storm-water infiltration and for constructing treatment systems and interceptors.[46] The EPA estimated that an additional $23.2 billion would be needed to correct pollution problems resulting from sewer overflows. Thus, total costs of meeting the needs would be substantially greater than the total spending possible under the most generous of the federal cost-sharing proposals considered by Congress.

Three main questions must be addressed in any evaluation of this federal subsidy program. The first is a multiple-part question: Should the

Table 4–11. Comparison of Proposed Funding Provisions for the Municipal Construction Grant Program (in billions of dollars)

	Office of Management and Budget	House of Representatives: HR-8	Senate: S-1128
Total federal grant authorization	$ 6.0 for 1986–89	$12.0 for 1986–88	$ 9.6 for 1986–90
Total spending enabled (at 55% federal cost share)	$10.9	$21.8	$17.5
Federal contribution to state revolving loan funds	$ 0.0	$ 9.0 for 1986–94	$ 8.4 for 1991–94
Total federal contributions	$ 8.4	$21.0	$18.0

Sources: "Clean Water Debate to Focus on Sewage Grant Program," *Congressional Quarterly*, March 16, 1985, pp. 491–492; "Clean Water Act Amendments of 1985," Committee on Environment and Public Works, U.S. Senate, Report 99-50, 99 Cong. 2 sess. (May 14, 1985) and "Clean Water Act Renewal Awaits House, Senate Action," *Congressional Quarterly*, May 25, 1985, pp. 1009–1010.

federal government be contributing financially to the construction of municipal facilities, and if so at what level? Should the cost-sharing program be eliminated or phased down, maintained at more or less existing levels, or substantially increased to match estimated needs? If continuing federal involvement is deemed appropriate, the second main question is: What form should federal aid take? Options include outright grants, loans, and grants to state-managed revolving loan funds. The third main question concerns the allocation of federal aid: How can this aid be allocated so as to maximize the improvement in water quality?

Before taking up these questions, it will be useful to look at data on what has been accomplished already. According to the Council on Environmental Quality, between 1973 and 1983 the total authorization for the federal construction grant program came to $43.8 billion. Through 1982, $34 billion of this had been obligated and $25 billion actually spent. This federal spending, combined with the municipalities' contributions, had resulted in the completion of 9,000 sewage treatment plants and was supporting the completion of another 8,000 facilities.[47] The progress made between 1960 and 1982 in increasing the treatment of municipal waste water is shown in table 4–12. Despite an increase of 58 million in

Table 4–12. Population Served by Municipal Waste Water Systems, with Level of Treatment

	1960	1970	1978	1980	1982
Total U.S. Population (in millions)	180	203	220	224	230
Population served by waste water system (in millions)	110	145	154	159	168
Percentage of population	61%	71%	70%	71%	73%
Percentage of served population with:					
no treatment	63%	41%	1%	1%	8%
primary treatment	33	NA	31	26	22
at least secondary treatment	4	NA	68	73	69

Note: NA = not available.
Source: Council on Environmental Quality, *Environmental Quality—1982* (Washington, D.C., 1982) table A-61, p. 295.

the population living in homes connected to municipal waste water systems (reflecting primarily the trend of urbanization), the percentage of the population whose wastes received no treatment fell from 63 percent to 8 percent. And the proportion of the population whose wastes received at least secondary treatment (a standard imposed by FWPCA-72) rose from only 4 percent in 1960 to 69 percent in 1982.

While the progress in increasing the treatment of municipal wastes was associated with increased federal financial aid going back to 1956, it is not possible to conclude that federal aid was fully responsible for the improvement. Available evidence suggests that to some extent municipalities used federal dollars to reduce their own financial burden for constructing facilities that would have been built in any event. Some evidence of this displacement of municipal spending by federal cost-sharing is indicated in figure 4–1. In constant dollars, state and local spending peaked in 1972 and had declined by more than 50 percent by 1982. One econometric study of federal and municipal spending estimated that each additional dollar of federal spending resulted in a reduction of municipal spending by 67 cents.[48]

One possible justification for a continuing federal role in financing the construction of municipal treatment facilities is that such financing is

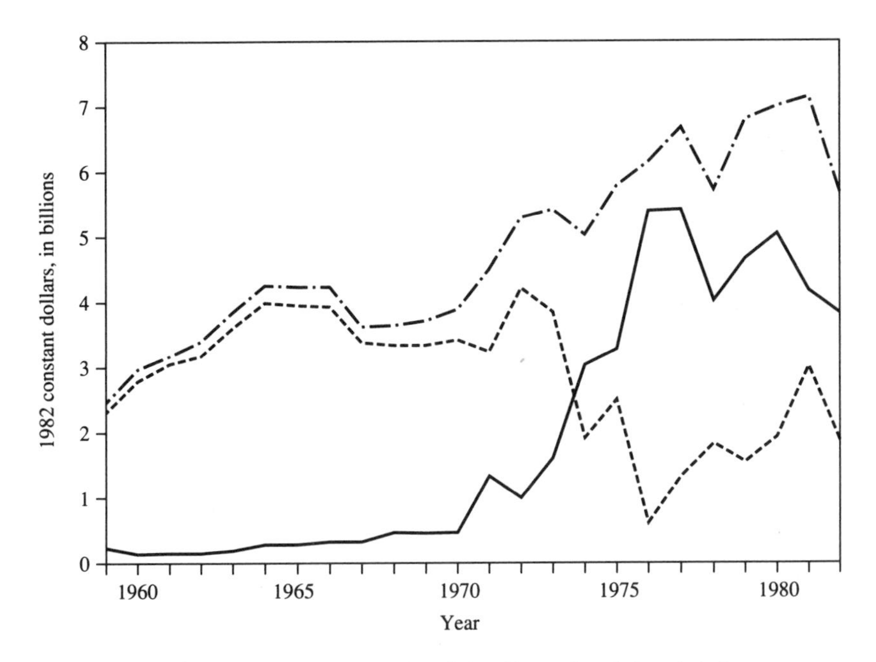

Figure 4–1. Substitution of state/local funding with federal dollars. *Source:* Environmental Protection Agency, *Study of the Future Federal Role in Municipal Wastewater Treatment—Report to the Administrator* (Washington, D.C., 1984) p. 3–2.

necessary to induce municipalities to comply with federal regulations requiring secondary treatment. But the evidence on displacement does not support this justification. It appears that the stick of federal regulation is relatively more important than the carrot of federal aid in influencing decisions about municipal pollution control. If this is the case, the question of the level and form of federal financial assistance is essentially a political one having substantial equity implications.

The federal construction grant program influences who pays for pollution control in two ways. The first is by shifting part of the cost away from user fees and state and local tax systems and on to the federal tax system. One study suggests that this shift could result in a fairer (or less regressive) distribution of the incidence of controlling municipal pollution.[49] It found that the costs to municipalities of meeting the requirements of FWPCA-72 (excluding runoff controls) imposed a burden equal to 0.6 percent of family income in the lowest income category but cost families in the highest income category only 0.3 percent of their income. In contrast, the federal cost-sharing program cost the lowest income families 0.4 percent of their income, with the incidence rising to 0.8 percent for the highest-income families.

The federal construction grant program also influences the distribution of costs through the formula for allocating federal funds among the states. A complex formula based on each state's estimated need for new treatment facilities is the basis for this allocation. The effect of the federal grant program is to shift the financial burden toward states with high federal tax bases and low estimated needs and away from low-income states with high estimated needs.

The distributional consequences of federal aid can also be influenced by the form the aid takes. Each of the forms of aid to be considered here yields different distributions of the economic burden of pollution control and creates different sets of incentives to muncipal decision makers.

At present, federal aid takes the form of a grant to the municipality for a stated percentage of the design and construction costs for the facility. Many states provide additional grants to municipalities to cover a portion of the nonfederal share. As a consequence, many communities find themselves responsible for only 10 to 25 percent of the total construction cost. These communities have only weak incentives to hold the line on capital costs by seeking cost-effective design and technologies or by matching more carefully the designed capacity of the plant to projected need. Thus capital expenditures for federally aided plants are likely to be higher than necessary to achieve the required secondary treatment. One study has estimated that by substantially increasing the local share of costs, capital costs could be reduced by as much as 30 percent.[50] Any favorable redistributional effects associated with the federal construction grant program appear to come at the expense of significant inefficiencies and excess costs.

The Water Quality Act of 1987 calls for phasing out the federal construction grant program by 1991 and replacing it with federal contributions to state-managed revolving loan funds. To meet their share of local costs, municipalities would be able to borrow directly from the state loan funds (presumably at a below-market interest rate) rather than compete for funds in private capital markets. The greater local burden associated with the reduced federal share of costs would be only partially offset by the reduction in excess capital cost and inefficiency and the lower borrowing costs. It is therefore understandable that municipalities favor the continuation of the federal grant program over a revolving loan fund.

Another form that federal aid could take is a subsidy to operating and maintanence costs rather than to construction costs. But subsidies in this form could also distort incentives and lead to inefficiencies or ineffective plant operation. For example, if the federal government paid a percentage of actual operating and maintenance costs, plant operators would have weaker incentives to avoid waste and operate their plants efficiently.[51]

The third main question to be discussed is how to get the best results in improved water quality from federal aid to municipalities. One recommendation stems from the gap between the number of projects that can be supported from likely federal funding levels and estimates of municipal needs. Clearly not all eligible projects can be supported by federal funds, so priority should be given to aiding those projects which will have the greatest impact on water quality. A recent study by the EPA found that more than half of the proposed projects identified in a 1982 Needs Survey were located on water bodies which already were meeting standards for designated uses.[52] Marginal benefits for completing these projects would be relatively small.

A mechanism exists in the federal law for implementing such a priority system: it is the requirement for areawide waste water management planning stipulated in section 208 of FWPCA-72. The section 208 planning process was not given high importance by the EPA in the early stages of implementing the 1972 law. Yet from a resource management point of view, it is important to maintain—and where necessary strengthen—the planning process with a particular eye toward using it to identify the highest-priority projects for federal cost sharing.

A second suggestion for maximizing the improvements achieved with federal aid is to seek more cost-effectiveness in allocating funds to specific categories of projects. At present, projects to correct problems of overflows from combined sewer systems and other sources of urban non-point pollution are not eligible for federal funding. But the EPA has found that in some river basins and for some substances (especially total suspended solids and lead), correcting combined sewer overflow problems and controlling other urban non-point sources may be substantially more cost-effective in meeting water quality standards than increasing control at municipal treatment facilities.[53] As the EPA report states,

> The Clean Water Act and its amendments have discouraged funding of projects that control urban nonpoint pollution sources . . . However, if these nonpoint source loadings are significant, then more rather than less discretion is needed to meeting [*sic*] the objectives of the Act . . . The results [of this analysis] serve to focus attention on the economic efficiency of targeting funds at programs that achieve water quality improvements. It further suggests that funding some, but not all, of the combined sewer overflow projects can be justified on an economic benefit-cost basis.[54]

One problem of continuing concern should be pointed out here—the problem of effective operation of municipal treatment plants to achieve designed standards of waste removal.[55] Although EPA officials have testified before Congress that compliance rates of municipal treatment plants are high (around 80 percent),[56] this testimony is not consistent with evidence of significant noncompliance obtained by the General Accounting Office.[57] If cost-conscious municipalities are too frugal in their operating and maintenance expenditures, they may fail to achieve effluent limitations and water quality targets.

The absence of adequate incentives for proper operation and maintenance is one cause of this problem. A possible solution would be a system of effluent charges. An effluent charge based on actual discharges would increase the financial incentive to operate treatment plants effectively. It would also increase cost-effectiveness in achieving reductions in total quantities discharged, especially if it were combined with effluent charges on industrial sources. A state-managed system of effluent charges could be made essentially revenue-neutral by returning funds to the municipalities through grants or subsidies as long as the rules for returning funds were independent of the actual levels of discharges at treatment plants.

Non-point Sources of Pollution

Several features of non-point source pollution must be taken into account in any discussion and evaluation of policy options. Among them is the wide variety of sources of such pollution. It can come from urban storm-sewer systems that discharge directly to water bodies, from urban runoff into streams and drainage ditches, from non-urban sources such as agricultural and silvicultural runoff from erosion, and from surface and subsurface mining activities. The variety of sources is associated with a variety of pollutants, including oxygen-demanding organic material, fecal bacteria, suspended solids and sediments, heavy metals, toxic organic compounds, and two plant nutrients—phosphorus and nitrogen—which can degrade water quality when present in excessive amounts. The variety of sources and pollutants also means that a variety of control technologies may be required. These may range from the interception, collection, and treatment of runoff waters to remove pollutants all the way to changes in agricultural, forestry, and mining practices to prevent runoff or prevent pollutants from being carried away by normal runoff. Another feature of non-point source pollution is its episodic character: it occurs primarily in connection with periods of rainfall or the melting of accumulated snow. This is in contrast to the more-or-less steady flow of pollutants from municipal sewer systems and many industrial processes.

It has now become clear that some degree of control of non-point source pollution will be essential if water quality standards are to be achieved throughout the country. In 1981, thirty-seven states reported that pollution from non-point sources would prevent their attaining the national objective of fishable and swimmable water quality.[58] This conclusion is supported by a recent water quality modeling study (discussed earlier in the section on Changes in Water Quality), which analyzed the effects of three different control programs on four water quality parameters—phosphorus, nitrogen, biochemical oxygen demand, and dissolved oxygen.[59] The three programs were BPT and ST at point sources, elimination of non-point discharges from all nonirrigated cropland, and a combination of these two. As indicated in table 4–13, for each of the pollutants control of sediment from non-point sources results in a bigger improvement in water quality (as measured by the number of stations meeting standards) than does BPT/ST alone. The difference is substantial in the case of the two nutrients. A comparison of a combined policy of BPT/ST and BAT/BCP with sediment control would be more relevant, but given the overwhelming contribution of non-point sources to total discharges of phosphorus

Table 4–13. Effects of BPT/ST, Agricultural Sediment Controls, and Combined Controls on Water Quality

	Percentage of locations meeting assumed water quality standards			
	1972 baseline	BPT/ST	Agricultural sediment control	Both controls
Total phosphorus[a]	27	32	39	48
Total Kjeldahl nitrogen[b]	30	32	43	48
Biochemical oxygen demand[c]	68	75	76	83
Dissolved oxygen[d]	83	88	89	93

Source: Gianessi and Peskin, "Analysis of National Water Pollution Control Policies," pp. 805–807.

[a] The standard is total phosphorus $\leq .180$ mg/l.

[b] The standard is total Kjeldahl nitrogen $\leq .90$ mg/l.

[c] The standard is 5-day biochemical oxygen demand ≤ 9.0 mg/l.

[d] The standard is dissolved oxygen > 5.0 mg/l.

and nitrogen, it is unlikely that the inclusion of BAT/BCP would alter the conclusions regarding nutrients.

Even in those water bodies where significant control of non-point source pollution is not required for attaining water quality standards, it may be desirable on grounds of cost-effectiveness to control it anyway. Wherever the incremental costs of non-point source control are less than the costs of controlling point sources to meet the BAT/BCP standards, the total cost of meeting water quality standards could be reduced by relaxing the control requirements on point sources and substituting the relatively less costly control of non-point sources of the same pollutants. Relatively little analysis of such a policy has been carried out to date. But the EPA study of urban point and non-point municipal sources of pollution discussed above shows that there may be options for making tradeoffs between municipal point and non-point sources.[60] If this is the case, it should also be possible to reduce total costs even further by trading off between municipal non-point sources and industrial point sources.

A case study of the Dillon reservoir in the Colorado Rockies confirms the potential for cost savings by trading non-point and point source pollution control requirements.[61] Four municipal sewer systems constitute the only major point sources of pollution to the reservoir. The total of point and non-point source discharges of phosphorus were projected to exceed the assimilative capacity of the reservoir. In comparison with more stringent treatment of point source discharges, a policy of non-point source control was estimated to result in control cost savings of more than $1 million per year. The control of non-point sources of phosphorus involved the relatively simple expedient of intercepting the runoff, allowing solids to settle out, and passing the remaining effluent through a sand and rock filtration system.

Turning to the regulatory framework, the present federal policy for non-point source pollution control implies a belief that the problem is best dealt with by state and local agencies. Section 208 of FWPCA-72 requires state planning agencies to include consideration of non-point sources in developing areawide waste management plans, and the Water Quality Act of 1987 includes a new section detailing state responsibilities for control of non-point source pollution. But neither provision establishes authority for federal regulation.

In one respect the decision to leave non-point source control planning to the state and local agencies may be a virtue. If Congress had followed the same approaches as it did with point sources—that is, reliance on technology-based effluent standards—we might now be saddled with an inflexible and highly costly set of national uniform design requirements for the control of non-point sources. As a recent EPA report put it,

> The basic approach taken by the Clean Water Act for managing point sources —that is, the application of uniform technological controls to classes of dischargers—is not appropriate for the management of nonpoint sources. Flexible, site-specific, and source-specific decision making is the key to effective control of nonpoint sources. Site-specific decisions must consider the nature of the watershed, the nature of the water body, the nature of the nonpoint source(s), the use impairment caused by the nonpoint source(s), and the range of management practices available to control nonpoint source pollution. The actual site-specific selection of particular management practices to control nonpoint source pollution (called "Best Management Practices") will involve local environmental and economic considerations, as well as considerations of effectiveness and acceptability of the practice.[62]

On the other hand, if non-point sources in one state contribute significantly to pollution problems in other states farther downstream, federal involvement in planning and regulations may be required. It should be recalled that the interstate character of many of the water pollution problems in the 1950s and 1960s was one factor leading to increasing involvement of the federal government in regulating point sources of pollution. The Water Quality Act of 1987 takes a leaf from the 1956 law in authorizing interstate management conferences to deal with cases of transboundary non-point source pollution.

The Water Quality Act of 1987 also requires that states develop non-point source pollution control programs specifying which best management practices are required of parties responsible for non-point source pollution. The act provides for federal grants to share in the states' costs of implementing their management programs, and encourages states to require best management practices only where non-point source pollution makes a significant contribution to ambient water quality problems. The act also calls for examining potential impacts of best management practices on groundwater quality during the development of the control program, but does not establish specific clean-up targets or deadlines for implementation. Clearly needed are more federal leadership and stronger incentives to state agencies to implement those non-point source control options identified through the mandated planning process. The flexibility necessary to take account of the heterogeneous nature of non-point source problems would thus be preserved.[63]

Water quality can be affected for better or for worse by policies not directed specifically at water pollution control. For example, the federal agricultural price support program has had the effect of encouraging farmers to bring marginal lands into cultivation, thus increasing erosion and non-point source pollution problems. By the same token, the federal Conservation Reserve Program, enacted in 1985, appears to be having a beneficial effect on water quality. The Conservation Reserve Program has the objective of removing some 40 to 45 million acres of highly erodable land from cultivation by subsidizing farmers' costs of planting grass or trees and contributing annual rental payments. The U.S. Department of Agriculture has estimated that among the consequences of the program will be a reduction of total erosion in the United States by 8.6 percent, and a reduction in nitrogen, phosphorus, and total suspended solids (TSS) loadings by between 7.8 and 9.6 percent nationwide. The present value of benefits due to improved water quality is estimated to lie in the range of $1.9 to $5.5 billion.[64]

CONCLUSIONS

Three major themes can be traced through this discussion of water pollution control policy. They are the importance of comparing benefits and costs, the value of seeking more cost-effective control programs, and the potential role for economic incentives such as charges or marketable discharge permits.

We have seen that in aggregate it appears that the costs of the present policy substantially outweigh the benefits. Yet if the goal of fishable and swimmable water quality is to be met everywhere, even more costs will have to be incurred. If it is accepted that the resources presently devoted to water pollution control are scarce, involve opportunity costs, and may have more valuable uses in other activities, a reconsideration of some water pollution goals and standards on a case-by-case basis might be necessary. This may mean accepting less than fishable-swimmable water quality where the costs of obtaining it are inordinately high. And it could mean requiring less than the best available technology or the best conventional control technology where already high water quality means that additional control brings very few additional benefits.

It has been argued here that one way to improve the benefit-cost relationship in the existing policy is to seek more cost-effective means of achieving given standards. The emphasis on equal treatment of dischargers or uniformity of clean-up requirements has meant that the cost of reaching present water quality objectives is substantially higher than necessary. Thus fewer of society's resources are available for other valuable uses. More emphasis should be given to the development of cost-effective means of achieving targets. We have discussed the potential role of charges or marketable discharge permits in moving toward a more cost-effective pollution control policy. Yet even if charge strategies are not adopted, substantial savings could be realized through more selective rather than uniform application of discharge standards such as BAT, and by considering tradeoffs between controls on point and non-point sources discharging to the same water body.

Finally, we have seen that progress toward attaining water pollution control objectives has been slow. Timetables have not been kept, and deadlines have been reached and passed without full compliance with the legislated objectives. These shortfalls in implementation are due in substantial part to the complexities of the task. But a major share of the responsibility for the slow pace of progress must be assigned to the inappropriate incentive structures created by the regulatory approach to pollution control. There are opportunities for restructuring incentives through the imposition of various types of charges for pollution. We have suggested marketable discharge permits or charges on all pollution dischargers to strengthen the incentives for proper operation of treatment systems and for innovation and technological change in methods of reducing pollution.

NOTES

1. David Zwick and Marcy Benstock, *Water Wasteland: Ralph Nader's Study Group Report on Water Pollution* (New York, Grossman Publishers, 1971).

2. See National Commission on Water Quality, *Report to the Congress* (Washington, D.C., March 1976); and National Commission on Water Quality, *Staff Report* (Washington, D.C., March 1976). The 1977 amendments and some of the findings of the commission are discussed in A. Myrick Freeman III, "Air and Water Pollution Policy," in Paul R. Portney, ed., *Current Issues in U.S. Environmental Policy* (Baltimore, The Johns Hopkins University Press for Resources for the Future, 1978) pp. 45–65.

3. National Commission on Water Quality, *Report to the Congress*, pp. 285–286.

4. For accounts of a short-lived effort to use the Refuse Act as the basis for a national water pollution control program in the early 1970s, see ibid., pp. 285–301; and J. Clarence Davies III and Barbara S. Davies, *The Politics of Pollution* (2nd ed., New York, Bobbs-Merrill, 1975) pp. 208–210.

5. Much of the following discussion of the history of federal policy is drawn from Allen V. Kneese and Charles L. Schultze, *Pollution, Prices, and Public Policy* (Washington, D.C., Brookings Institution, 1975) pp. 30–45.

6. See Zwick and Benstock, *Water Wasteland*, pp. 264–284.

7. See Allen V. Kneese and Blair T. Bower, *Managing Water Quality: Economics, Technology, Institutions* (Baltimore, The Johns Hopkins University Press for Resources for the Future, 1968) chap. 11.

8. See T. H. Tietenberg, *Environmental and Natural Resource Economics* (Glenview, Ill., Scott, Foresman and Co., 1984) p. 352.

9. See 33 U.S.C. 466, sections 301 (g)(1)(C) and 301 (h)(2).

10. For further discussion of goals and the relationship between means and goals, see Freeman, "Air and Water Pollution Policy," pp. 48–53.

11. For further discussion of technology-based standards, see Freeman, "Air and Water Pollution Policy," pp. 53–58; and A. Myrick Freeman III, "Technology-Based Standards: The U.S. Case," *Water Resources Research* vol. 16, no. 1 (February 1980) pp. 21–27.

12. Information provided to the author by the Environmental Protection Agency, March 1989.

13. National Commission on Water Quality, *Staff Report,* p. V-34.

14. For a description of the procedures for issuing effluent limitations regulations and the process used by the EPA in developing these regulations, see Wesley A. Magat, Alan J. Krupnick, and Winston Harrington, *Rules in the Making: A Statistical Analysis of Regulatory Agency Behavior* (Washington, D.C., Resources for the Future, 1986) chap. 3.

15. Information provided to the author by the EPA, March 1989.

16. Council on Environmental Quality, *Environmental Quality—1976* (Washington, D.C., 1976) p. 15.

17. General Accounting Office, *Waste Water Dischargers Are Not Complying with EPA Pollution Control Permits* (Washington, D.C., 1983) p. 2.

18. Council on Environmental Quality, *Environmental Quality—1982* (Washington, D.C., 1982) p. 86.

19. Ibid., pp. 85–86.

20. General Accounting Office, *Waste Water Dischargers,* pp. 13–14.

21. See Council on Environmental Quality, *Environmental Quality* (Washington, D.C., 1974) pp. 280–289; and Council on Environmental Quality, *Environmental Quality* (Washington, D.C., 1975) pp. 350–355.

22. Leonard P. Gianessi, Henry M. Peskin, and G. K. Young, "Analysis of Water Pollution Control Policies: 1. A National Network Model," *Water Resources Research* vol. 17, no. 4 (August 1981) pp. 796–802; Leonard P. Gianessi and Henry M. Peskin, "Analysis of National Water Pollution Control Policies: 2. Agricultural Sediment Control," *Water Resources Research* vol. 17, no. 4 (August 1981) pp. 803–821.

23. William J. Vaughan and Clifford S. Russell, *Freshwater Recreational Fishing: The National Benefits of Water Pollution Control* (Washington, D.C., Resources for the Future, 1982) pp. 38–48.

24. Ibid., table 2-5.

25. These results are described in The Conservation Foundation, *State of the Environment: An Assessment at Mid-Decade* (Washington, D.C., The Conservation Foundation, 1984) pp. 106–109.

26. Richard A. Smith, Richard B. Alexander, and M. Gordon Wolman, "Water-Quality Trends in the Nation's Rivers," *Science* vol. 235 (March 22, 1987) pp. 1607–1615.

27. For a clear and concise introduction to the methods of estimating pollution control benefits, see Allen V. Kneese, *Measuring the Benefits of Clean Air and Water* (Washington, D.C., Resources for the Future, 1984). The more technically oriented reader can find descriptions of the relevant economic models in A. Myrick Freeman III, *The Benefits of Environmental Improvement: Theory and Practice* (Baltimore, The Johns Hopkins University Press for Resources for the Future, 1979); and A. Myrick Freeman III, "Methods for Assessing the Benefits of Environmental Programs," in Allen V. Kneese and James L. Sweeney, eds., *Handbook of Natural Resource and Energy Economics,* vol. 1 (Amsterdam, North-Holland, 1985).

28. A. Myrick Freeman III, "The Benefits of Air and Water Pollution Control: A Review and Synthesis," Report to the Council on Environmental Quality, Executive Office of the President, December 1979; updated and revised as Freeman, *Air and Water Pollution Control: A Benefit-Cost Assessment* (New York, Wiley, 1982).

29. See Council on Environmental Quality, *Environmental Quality—1980* (Washington, D.C., 1980). A useful discussion of this set of cost estimates and the general problem of defining and estimating true pollution control costs can be found in Paul R. Portney, "The Macroeconomic Impacts of Federal Regulation," in Henry M. Peskin, Paul R. Portney, and Allen V. Kneese, *Environmental Regulation and the U.S. Economy* (Baltimore, The Johns Hopkins University Press for Resources for the Future, 1981).

30. Environmental Protection Agency, *Final Report: The Cost of Clean Air and Water*, Report to Congress, 1984 (Washington, D.C., May 1984).

31. See, for example, Portney, "Macroeconomic Impacts."

32. National Commission on Water Quality, *Report to the Congress*, p. 21.

33. 33 U.S.C. 466, section 304(b)(4)(A).

34. In a personal communication to the author, July 26, 1985, Henry M. Peskin suggested that in practice there may be more latitude to adjust permit terms to particular circumstances. He writes, "Our cursory review of permit files seems to indicate quite a lot of flexibility on the part of the permit writer. In fact, sometimes it is hard to see any relationship between permits and [effluent limitation] guidelines."

35. For a discussion of this type of innovation to policy, with some estimates of potential cost savings, see Donna Downing and Stuart Sessions, "Innovative Water Quality-Based Permitting: A Policy Perspective," *Journal of the Water Pollution Control Federation* vol. 57, no. 5 (May 1985) pp. 358–365.

36. Magat, Krupnick, and Harrington, *Rules in the Making*, table 6–1.

37. Ralph A. Luken, Daniel J. Basta, and Edward H. Pechan, *The National Residuals Discharge Inventory* (Washington, D.C., National Research Council, 1976) chap. 9. Earlier studies comparing the costs of uniform treatment policies with cost-minimizing alternatives found the former to be sometimes two to three times larger than the latter. See also Kneese and Bower, *Managing Water Quality*, pp. 158–164, 224–235.

38. Allen V. Kneese and Charles L. Schultze are perhaps the best-known advocates of greater reliance on charges and other economic incentives; see Kneese and Schultze, *Pollution, Prices, and Public Policy*. See also Frederick R. Anderson, Allen V. Kneese, Phillip D. Reed, Serge Taylor, and Russell B. Stevenson, *Environmental Improvement Through Economic Incentives* (Baltimore, The Johns Hopkins University Press for Resources for the Future, 1977). For a more critical perspective, see Susan Rose-Ackerman, "Effluent Charges: A Critique," *Canadian Journal of Economics* vol. 6 (1973) pp. 512–528; Susan Rose-Ackerman, "Market Models for Pollution Control: Their Strengths and Weaknesses," *Public Policy* vol. 25, no. 3 (1977) pp. 383–406; and Clifford S. Russell, "What Can We Get from Effluent Charges," *Policy Analysis* vol. 5, no. 2 (Spring 1979) pp. 155–180.

39. See Kneese and Bower, *Managing Water Quality*, pp. 158–164; Henry W. Herzog, Jr., "Economic Efficiency and Equity in Water Quality Control: Effluent Taxes and Information Requirements," *Journal of Environmental Economics and Management* vol. 2, no. 3 (February 1976) pp. 170–184.

40. See Gardner M. Brown, Jr. and Ralph W. Johnson, "Pollution Control by Effluent Charges: It Works in the Federal Republic of Germany, Why Not in the United States," *Natural Resources Journal* vol. 24, no. 4 (October, 1984) pp. 929–966. For discussion of other European policies that incorporate effluent charges in some form, see Ralph W. Johnson and

Gardner M. Brown, Jr., *Cleaning Up Europe's Waters: Economics, Management, and Policies* (New York, Praeger, 1976); and Blair T. Bower, Rémi Barré, Jochen Kühner, and Clifford S. Russell, with Anne J. Price, *Incentives in Water Quality Management: France and the Ruhr Area* (Washington, D.C., Resources for the Future, 1981).

41. William B. O'Neill, Martin H. David, Christina Moore, and Erhard F. Joeres, "Transferable Discharge Permits and Economic Efficiency: The Fox River," *Journal of Environmental Economics and Management* vol. 10, no. 4 (December 1983) pp. 346–355. For a more general discussion of permit trading, see Erhard F. Joeres and Martin H. David, eds., *Buying a Better Environment: Cost-Effective Regulation Through Permit Trading* (Madison, University of Wisconsin Press, 1983). A more technical treatment and application primarily to the control of air pollution is in T. H. Tietenberg, *Emissions Trading: An Exercise in Reforming Pollution Policy* (Washington, D.C., Resources for the Future, 1985).

42. The bubble concept, first applied to multiple-stack sources of air pollution, was given this name because it treats a collection of stacks or sources as if they were encased in a bubble. Pollution control requirements are applied to the aggregate of emissions leaving the bubble rather than to each individual stack.

43. See Robert C. Greene, "Water Pollution Control for the Iron and Steel Industry," in James C. Miller III and Bruce Yandle, eds., *Benefit-Cost Analysis of Social Regulations* (Washington, D.C., American Enterprise Institute, 1979).

44. See Water Quality Committee, "Annual Review of Significant Activities—1983," *Natural Resources Lawyer* vol. 17, no. 2 (1983) pp. 277–278.

45. "Clean Water Debate to Focus on Sewage Grant Program," *Congressional Quarterly*, March 16, 1985, pp. 491–492.

46. Environmental Protection Agency, *Assessment of Needed Publicly Owned Waste Water Treatment Facilities in the United States: 1984 Needs Survey Report to Congress* (Washington, D.C., February 1985) p. 8.

47. Council on Environmental Quality, *Environmental Quality—1982*, pp. 83–84.

48. James Jondrow and Robert A. Levy, "The Displacement of Local Spending for Pollution Control by Federal Construction Grants," *American Economic Review* vol. 74, no. 2 (May 1984) pp. 174–178.

49. Leonard P. Gianessi and Henry M. Peskin, "The Distribution of the Cost of Federal Water Pollution Control Policy," *Land Economics* vol. 56, no. 1 (February 1980) pp. 85–102.

50. U.S. Congress, Congressional Budget Office, *Efficient Investments in Wastewater Treatment Plants* (Washington, D.C., 1985).

51. For further discussion of this issue, see Freeman, "Air and Water Pollution Policy," pp. 61–63.

52. Environmental Protection Agency, Office of Policy Analysis, "Focusing on Water Quality Issues in Urban Areas: Loadings, Control Costs, and Water Quality Impacts of Point and Non-Point Municipal Sources of Pollution," draft report, September 27, 1984.

53. Ibid., sections II and III.

54. Ibid., pp. 1–1, 1–4.

55. This problem is extensively discussed elsewhere. See, for example, Freeman, "Air and Water Pollution Policy," pp. 62–63.

56. Council on Environmental Quality, *Environmental Quality—1981* (Washington, D.C., 1981) pp. 75–77.

57. General Accounting Office, *Waste Water Dischargers*.

58. Council on Environmental Quality, *Environmental Quality—1981*, pp. 71–72.

59. Gianessi and Peskin, "Analysis of National Water Pollution Control Policies."

60. Environmental Protection Agency, Office of Policy Analysis, "Focusing on Water Quality Issues." See also Environmental Protection Agency, *Non-point Source Pollution in the U.S., Report to Congress* (Washington, D.C., 1984), for an explicit recognition of the potential for improved cost-effectiveness through selected targeting of high-payoff non-point source problems.

61. Industrial Economics, Inc., *Case Studies on the Trading of Effluent Loads: Dillon Reservoir, Final Report* (Cambridge, Mass., Industrial Economics, 1984).

62. Environmental Protection Agency, *Non-Point Source Pollution in the U.S.*, pp. xiii–xiv.

63. For a useful discussion of policy options in the context of agricultural non-point source pollution, see Winston Harrington, Alan J. Krupnick, and Henry M. Peskin, "Policies for Nonpoint Source Water Pollution Control," *Journal of Soil and Water Conservation* vol. 40, no. 1 (January-February 1985) pp. 27–32.

64. Marc O. Ribaudo, *Water Quality Benefits from the Conservation Reserve Program*, Agricultural Economic Report no. 606 (Washington, D.C., Economic Research Service, U.S. Department of Agriculture, 1989).

five

Hazardous Wastes

Roger C. Dower

INTRODUCTION

Throughout the 1970s air and water pollution control was the main focus of U.S. environmental improvement efforts. While much was written during that period on the need to take a "materials balance" approach to environmental policy—that is, to recognize that what goes in must come out—the disposal of solid and liquid wastes on land received relatively little attention. Viewed primarily as a local or regional problem of littering and trash removal, solid waste disposal was seen as secondary to pressing public health and amenity issues associated with air and water pollution.

One can almost mark 1978 as the year when public attention shifted radically toward the view of hazardous waste disposal as a national environmental problem. It was then that the nation first learned of Love Canal, a residential area of Niagara Falls, New York, where large quantities of solid and liquid wastes, long buried under ground where a school and an adjacent housing tract were later built, had begun to seep into the basements, playrooms, and general environment of area households. Fears of serious adverse health effects began to grow, and extended to other parts of the country as similar sites began to receive attention. Today, no other environmental problem is more well-publicized or higher on the public agenda than hazardous wastes; more than 60 percent of the respondents in a recent poll indicated that they view the problem as "very serious."[1] In 1989, the Environmental Protection Agency alone will spend over $1.4 billion to deal with the disposal of solid wastes.[2]

The subject of this chapter is that subset of solid and liquid wastes that is at the center of the current regulatory storm: hazardous wastes. The definition of hazardous wastes is not entirely clear. We tend to think of hazardous wastes as being disposed of on land, although many are not, and we worry about them because of their threats to biological systems (in contrast to our concerns over the more aesthetic types of damage associated with other solid wastes). As earlier chapters have explained, direct emissions of effluents into the air or water are regulated under the Clean Air and Clean Water acts. Similarly, direct or indirect contamination of drinking water supplies is addressed under the Safe Drinking Water Act. These acts leave for additional regulation the solid or liquid wastes that are disposed of on land, even if they ultimately result in air pollution (as they sometimes do) or the contamination of surface or ground waters (as they often do).

THE NATURE AND SCOPE OF THE PROBLEM

Given the amount of public attention accorded hazardous wastes, it is surprising how little is known about their nature and the scope of the problem. In this section, we examine definitions of hazardous wastes, outline the dimensions of the problem, and attempt to bring hazardous wastes into perspective as an environmental concern.

What Is Hazardous Waste?

As we have seen, hazardous wastes can be informally defined as a subset of all solid and liquid wastes, which are disposed of on land rather than being shunted directly into the air or water, and which have the potential to adversely affect human health and the environment. The tendency is to think of hazardous wastes as resulting mainly from industrial activities, but households also play a role in the generation and improper disposal of substances that might be considered hazardous wastes.

Under various statutes and regulations Congress and the EPA have defined hazardous wastes more carefully. Current law, for example, defines hazardous wastes as those solid or liquid wastes or combinations thereof that may "cause or contribute to an increase in mortality or an increase in serious irreversible, or incapacitating reversible illness" or any wastes which may "pose a substantial threat to human health" when improperly handled.[3] The Environmental Protection Agency has refined and complicated this definition in its own regulations.[4] But at the risk of severe oversimplification, the EPA has also adopted the convention that a hazardous waste is any solid or liquid waste (that is, any material that is thrown away) that has certain harmful effects on health or the environment.[5] The legal definition of hazardous wastes is not all-inclusive, however. Under current law and regulations, several categories of potential hazardous wastes are not regulated for political and other reasons. These include mining wastes, cement-kiln dust, household wastes, and agricultural wastes.

It should be noted that the focus on land disposal may not fully capture the nature of the problem. Any effluents not covered under the air and water statutes may be thought of as hazardous wastes—whether disposed of on land, directly discharged into the air or water, or placed in oceans or under the ground. The current statutory structure for hazardous waste management covers all of these disposal routes. In the very broadest sense, all pollutants might be considered hazardous wastes to the extent that they endanger public health or welfare. Our current regulatory approach to hazardous wastes is not so much predicated on the definition of hazardous wastes (although this is an important issue) as it is on the source of the wastes introduced into the environment.

What Is the Problem?

Broadly speaking, current concerns about hazardous wastes focus on the potential for improper storage or disposal to lead to environmental or human exposure. Wastes placed in plain metal drums can cause corrosion and leak out into the general environment. Wastes held in unlined ponds, lagoons, or landfills over long periods of time may leach into the surrounding soil and nearby water supplies. Our concern with hazardous wastes also focuses on two important time dimensions: past disposal practices have resulted in present risks to health and the environment; and some current disposal practices (such as placing wastes in unprotected landfills) may result in unacceptable levels of risk in the future. The distinction between past and present hazardous waste disposal practices is quite important. Programs and incentives directed at current disposal practices can have no effect on past activities. However, the design of programs for cleaning up and allocating the liability for past waste sites may influence, in unexpected directions, those currently generating and disposing of hazardous wastes.

The conditions that led to the creation of particular hazardous waste problems are the same as those that underlie most current environmental concerns. As long as emissions to the air, water, or land are free, little economic incentive exists for firms or households to find alternative, less damaging disposal options. As long as the price of waste disposal is artificially low (through failure to include the social or environmental costs associated with the activity), too many wastes are produced and too few go into safe disposal facilities. It is not difficult to see why firms and others faced with the costs of incineration to render wastes less harmful (estimated to range from $300 to $1,000 per ton) or of burying the wastes in landfills (perhaps as little as $50 per ton) would choose the latter.

The scenario is the same for air and water pollution, but other characteristics of hazardous wastes have exacerbated the problem relative to the more conventional pollution problems. First, it has been comparatively easy to hide poor hazardous waste disposal practices from the public—much easier than hiding smoke billowing from a stack or liquid effluent spewing from an outfall pipe. Landfills are located in out-of-the-way places and many are on privately held land. There may be few outward signs of disposal, safe or unsafe.

Second, hazardous wastes can affect all environmental media—air, water, and land. As noted above, wastes placed in metal barrels may begin over time to leak into the ground, and may slowly seep through the soil into underground aquifers or be carried through surface runoff into streams or rivers. Emissions from wastes stored above ground may mix with the surrounding air to pose a health threat to those downwind from or near the site. And with hazardous wastes the relevant linkages between the disposal of wastes and their ultimate effects on health and the environment can be much more numerous and complex than the linkages involved with conventional pollutants. Exposure routes are not often direct and may involve several different avenues simultaneously.

Third, a hazardous waste site may continue to pollute long after it has ceased accepting new wastes. One can stop an industrial facility from polluting the air by shutting down the plant. In a similar fashion, one can stop a plant's production of hazardous wastes. However, environmental and health risks may always be associated with a disposal facility even after it has closed. This characteristic has important implications for the design of effective regulatory and enforcement strategies, as we will see.

How Big Is the Problem?

While we know little about the actual scope of the hazardous waste problem, what we do know hints at its potential magnitude. According to a recent congressional study, the volume of hazardous wastes generated each year is in the neighborhood of 250 million metric tons.[6] These wastes are composed of a diverse mix of substances and come from almost every major industrial sector, as tables 5–1 and 5–2 show. There are an estimated 650,000 generators of hazardous wastes, but roughly 2 percent of them contribute over 95 percent of the wastes; most generators produce less than 1,000 kilograms per month. From the point of view of policy, it is important to note that of the quarter of a billion tons of hazardous wastes produced, about 95 percent are stored at the site at which they are generated.[7] The remaining 5 percent is shipped elsewhere for treatment and/or disposal.

Table 5–1. Estimated National Generation of Industrial Hazardous Wastes in 1983, Ranked by Waste Quantity (in thousands of metric tons)

Waste type	Estimated range		Mean quantity	Percentage of total
	Lower	Upper		
Nonmetallic inorganic liquids	68,102	96,420	82,261	31
Nonmetallic inorganic sludge	23,285	32,837	28,061	11
Nonmetallic inorganic dusts	19,455	22,784	21,120	8
Metal-containing liquids	14,125	25,394	19,760	7
Miscellaneous wastes	14,438	16,393	15,415	6
Metal-containing sludge	13,246	15,748	14,497	6
Waste oils	9,835	18,664	14,249	5
Nonhalogenated solvents	11,325	12,935	12,130	5
Halogenated organic solids	9,321	10,246	9,784	4
Metallic dusts and shavings	6,729	8,738	7,733	3
Cyanide and metal liquids	4,247	10,520	7,383	3
Contaminated clay, soil, and sand	5,092	5,839	5,461	2
Nonhalogenated organic solids	4,078	5,078	4,578	2
Dye and paint sludge	4,035	4,438	4,236	2
Resins, latex, and monomers	3,451	4,585	4,018	2
Oily sludge	2,965	4,502	3,734	1
Halogenated solvents	2,774	4,185	3,479	1
Other organic liquids	2,866	4,003	3,435	1
Nonhalogenated organic sludge	2,179	2,305	2,242	1
Explosives	508	933	720	[a]
Halogenated organic sludge	583	848	715	[a]
Cyanide and metal sludge	537	577	557	[a]
Pesticides, herbicides	19	33	26	[a]
Polychlorinated biphenols	1	1	1	[a]
Total	223,196	308,006	265,595	

Source: Congressional Budget Office, *Hazardous Waste Management: Recent Changes and Policy Alternatives* (Washington, D.C., Government Printing Office, 1985).

[a] Less than 1 percent.

Table 5–2. Estimated National Generation of Industrial Hazardous Wastes, Ranked by Major Industry Group (in thousands of metric tons)

Major industry	Estimated quantity in 1983	Percentage of total
Chemicals and allied products	127,245	47.9
Primary metals	47,704	18.0
Petroleum and coal products	31,358	11.8
Fabricated metal products	25,364	9.6
Rubber and plastic products	14,600	5.5
Miscellaneous manufacturing	5,614	2.1
Nonelectrical machinery	4,859	1.8
Transportation equipment	2,977	1.1
Motor freight transportation	2,160	0.8
Electrical and electronic machinery	1,929	0.7
Wood preserving	1,739	0.7
Drum reconditioners	45	—
Total	265,594	100.0

Source: Congressional Budget Office, *Hazardous Waste Management: Recent Changes and Policy Alternatives.*

Greater uncertainty surrounds the methods by which hazardous wastes are disposed of. Three very different estimates—by the Congressional Budget Office, the EPA, and the Chemical Manufacturers' Association—of the ultimate fate of hazardous wastes are shown in table 5–3. According to the Congressional Budget Office (CBO), the most common disposal method by volume is deep-well injection, where wastes are pumped into thin, deep shafts and allowed to collect beneath the earth's surface (sometimes in salt domes). In CBO's accounting, deep-well injection accounts for approximately 25 percent of all wastes disposed of off-site, another 23 percent goes into sanitary and hazardous waste landfills, and 5 percent is burned in incinerators. The rest are treated or disposed of through a variety of other techniques.

While the EPA and the Chemical Manufacturers' Association (CMA) paint a somewhat different picture, perhaps the most important single entry in table 5–3 is the large quantity of wastes discharged into surface waters—according to the CMA, more than 90 percent of the total. Such diffuse discharges are more difficult to identify and control than more contained discharges, and pose a more complicated management problem. Most of these discharges may be subject, however, to anticipated controls under the Clean Water Act.

Table 5–3. Estimates of Use of Different Hazardous Waste Management Methods

	Percentage of total		
Method	EPA	CBO	CMA[a]
Surface impoundment	55	19	<1
Injection into wells	13	25	8
Direct or indirect discharge to water	b	22	91
Landfill	1	23[c]	<1
Treatment	20	6	<1
Incineration	<1	5[d]	<1
Storage	9	b	b
Other	<1	b	b

Note: EPA = Environmental Protection Agency; CBO = Congressional Budget Office; CMA = Chemical Manufacturers' Association.

[a] Chemical industry only.

[b] Method not included in specific estimate.

[c] Includes hazardous waste disposed in sanitary landfills.

[d] Includes industrial boilers.

There is a tendency to assume that of the various disposal options incineration and treatment are the safest. Most of the current regulatory attention has therefore been focused on the wastes disposed of in landfills or storage tanks, or discharged directly into the environment. These would appear to be the disposal methods having the greatest chance of adversely affecting human health. There are currently operating in the United States about 500 licensed (permitted) commercial treatment, storage, and disposal facilities (TSDFs), 2,500 generator-owned TSDFs, and 75,000 industrial landfills. Most of the TSDFs are located in the Northeast and Midwest (see table 5–4). It is estimated that 10 generator-owned facilities account for more than 60 percent of all hazardous wastes managed.[8]

There is no accurate estimate of the number of formerly active landfills or other sites at which hazardous wastes were discarded in the past. The EPA currently lists 27,000 abandoned hazardous wastes sites on the inventory it maintains of sites that may require some sort of cleanup. Of these, the EPA estimates that cleanup of at least 2,000 will require some federal action. But the General Accounting Office has estimated that the number of sites may actually be anywhere from 130,000 to 425,000, depending on the definition of a hazardous waste site.[9] The Office of

Table 5–4. Hazardous Waste TSDFs and Waste Volumes, by Geographic Sector

Geographic sector	Number of TSD facilities	Quantity of RCRA waste managed (million metric tons)
Northeast	830	63
Southeast	440	84
Southwest	390	58
Midwest	910	64
Rocky Mountains	90	1
Far West	340	5
Total	3,000	275

Sources: Environmental Protection Agency, *National Screening Survey* (Office of Solid Waste, 1986); EPA, *The Hazardous Waste System* (Office of Solid Waste and Emergency Response, June 1987); and the EPA's Hazardous Waste Data Management System.

Technology Assessment (OTA) takes a broader and more pessimistic view, estimating that there are more than 600,000 active or former solid waste disposal facilities in the United States that could pose threats to health and the environment.[10] Not all of these would require federal attention to clean them up, but perhaps as many as 10,000 would, according to the OTA. If nothing else, these estimates reflect the range of uncertainty associated with past hazardous waste disposal; they do not hide the potential magnitude of the problem. None of the estimates, for example, include closed deep-well injection facilities, because of the difficulty of identifying them. Nor do these estimates include the number of potential hazardous waste sites owned by the federal government. Although federal agencies are only now beginning to assess the scope of their own problems, one recent study estimates that the number of federal facilities with hazardous waste problems is certainly over 1,000, as shown in table 5–5.[11]

What Are the Risks?

The mix of chemicals that typically constitutes the "soup" of hazardous wastes has been associated with a wide range of health and environmental effects, from acute disorders such as skin burns to chronic diseases such as cancer. The chemicals and compounds found commonly in dump sites with which the EPA is concerned, and the types of potential health risks associated with these chemicals and compounds, are listed in table 5–6.

Table 5–5. Number of Federal Facilities with Known Hazardous Waste Problems

Agency	No. of facilities
Department of Defense	572
Department of Energy	66
Department of Agriculture	39
Central Intelligence Agency	1
Department of Commerce	7
Environmental Protection Agency	17
General Services Administration	20
Department of Health and Human Services	4
Department of the Interior	263
Department of Justice	2
National Aeronautics and Space Administration	12
Postal Service	5
Small Business Administration	1
Tennessee Valley Authority	17
Department of Transportation	48
Department of Treasury	2
Veterans Administration	11
Total	1,087

Source: Congressional Budget Office, *Hazardous Waste Liabilities at Federal Facilities* (Washington, D.C., forthcoming).

Because of the wide variety of waste types, disposal sites, exposure conditions, and other important factors that exist, it is impossible to generalize about the health risks associated with an "average" site.

While the *potential* risks to health from exposure to hazardous wastes may be substantial, little is known about the *actual* risks to the public from past and current disposal practices. Very few studies or formal risk assessments have been conducted in the vicinity of abandoned or currently operating facilities. A recent study of 21 hazardous waste disposal sites found that in only one case was there strong evidence of adverse health effects associated with the site.[12] However, the study acknowledged that the findings were as much a statement on the inadequacy of the data base for making such assessments as they were on the apparent health effects.

Few defensible estimates of risks associated with current disposal practices are available, although the EPA has a number of studies under way. Those that do exist do not concern what is thought of as the biggest problem—land disposal. According to the EPA, emissions into the air of

Table 5–6. Typical Chemicals Found in Hazardous Waste Sites, and Their Potential Health Risks

Chemical	No. of sites (out of 546 sites)	Potential health risks
Trichlorethylene	179	Possible carcinogen
Lead	162	Acute toxicity in young children, associated with brain damage
Toluene	153	Carcinogen, possible neurotoxin
Benzene	143	Carcinogen
Polychlorinated biphenyls	121	Possible carcinogen, nervous/digestive disorders
Chloroform	111	Carcinogen, reproductive toxin

Source: Environmental Protection Agency, "Extent of the Hazardous Release Problem and Future Funding Needs—CERCLA Section 301(a)(1)(c) Study," December 1984.

volatile organic compounds from hazardous waste facilities *may* result in from 1 to 250 cases of cancer annually;[13] similarly, leachate and air emissions from burning, dumping, and disposal of used oil have been estimated to result in 80 annual cases of cancer nationwide.[14] Other estimates of health risks from hazardous waste sites are based on models that most observers feel may play a useful comparative role in standard-setting, but do not accurately reflect absolute risks. It is interesting, however, that a recent study conducted by the EPA's senior managers placed the risks from hazardous waste sites among the lowest the agency has to address.[15]

This cursory overview suggests several important dimensions of the hazardous waste problem that should be taken into account in any system designed to deal effectively with environmental and health concerns. First, uncertainty over risks suggests uncertainty over regulatory benefits. Second, for hazardous waste disposal—the last "unregulated" form of disposal—the costs of alternatives to land disposal are likely to be substantial. Together, these observations highlight the need for a control strategy that is flexible and designed to identify and act on the worst problems first. Cost and benefit information in the context of hazardous waste management has a high value. Thus, the incentives imbedded in the optimal hazardous waste programs have to serve several purposes: they should encourage the search for and implementation of least-cost solutions to current problems (including disposal, treatment, reduction, or recycling); at the same time, they should encourage the prompt and efficient cleanup of past disposal activities that pose current problems.

An economic-based approach to hazardous waste management would attempt to address these and other elements of the problem. However, the current statutory and regulatory system involves a mixed bag of incentives and constraints that call into question our ability to cope with hazardous wastes in an economically rational manner. Absent solid data or much in the way of a historical context, it is difficult to know whether we are going too far too fast, or whether we have undershot the mark; the following section outlines how the hazardous waste regulatory program has evolved and suggests that the former may be more likely than the latter.

THE CURRENT STATUTORY AND REGULATORY FRAMEWORK

The federal government has been concerned with the disposal and treatment of solid wastes since passage of the Solid Waste Disposal Act in 1965. Statutes and regulations concentrating on hazardous wastes are much more recent. It was not until 1976 that Congress created a regulatory program to deal with existing disposal practices, and it was not until 1980 that it came to grips with abandoned sites. In describing the statutory and regulatory structure that has emerged in response to the problem of hazardous wastes, we must caution that the task is akin to shooting at a moving target—the federal regulatory program is in a great deal of flux. Specifically, the 1976 law was entirely rewritten in 1984, and the regulations flowing from it are just beginning to appear and will continue to be issued into the early 1990s. The 1980 law concerning abandoned sites was reauthorized and significantly altered in 1986. However, certain basic characteristics of the programs have remained unchanged.

The Resource Conservation and Recovery Act (RCRA)

The basic federal statutory and regulatory framework for addressing current disposal practices and existing disposal sites emerged in 1976 as amendments to the Solid Waste Disposal Act. This set of amendments (and by now the act itself) is called the Resource Conservation and Recovery Act (or RCRA). The act was intended to provide a cradle-to-grave regulatory framework to monitor and control the production, storage, transportation, and eventual disposal of wastes that posed a risk to health and the environment. While Congress was clear on that goal, it was vague in spelling out the means to achieve it. Rather, it left the EPA with considerable flexibility and discretion in defining the ultimate approach. As remarked earlier, such flexibility would appear to be a particularly appropriate response to a problem as uncertain in breadth and seriousness as that of hazardous waste disposal practices. Yet this discretion may well have been the downfall of the program and to have led to a radical restructuring of RCRA by Congress in 1984.

The 1976 RCRA had four basic components relating to hazardous wastes. The first step required by the act was the identification and characterization of hazardous wastes. While the act contained general language concerning the definition of such wastes, it left the specifics to the EPA. The agency responded with a two-pronged approach. First, a list was developed of waste materials deemed hazardous on prima facie grounds and thus falling automatically under the regulatory requirements of the act. The EPA has now published a list of over 500 substances, chemical products, and mixtures that are considered hazardous.[16] Second, to determine whether any other wastes were to be deemed hazardous, the agency established four criteria—ignitability, corrosivity, reactivity, and toxicity.[17] Tests have been established for all four criteria. Should a waste prove "positive" on any one of these counts, it falls into the RCRA process, regardless of the importance of the products giving rise to the wastes. In that sense, the definitional process is risk-based and involves no balancing. (Certain large-volume wastes have been exempted to date, however, including those from oil and natural gas exploration.)

The second major thrust of RCRA (1976) dealt with the generation and transportation of wastes. Under the act, generators of hazardous wastes were given the responsibility of keeping track of the wastes they generate and where they go. Congress anticipated that the tracking function would include, at a minimum, the use of a system of manifests for following the movement of hazardous wastes from generation to disposal, but provided few additional details on what was expected. All generators, transporters, and disposers of hazardous wastes were required by the EPA to be a part of a system that would identify the quantity, origin, and destination of hazardous wastes being transported, as well as the identity of the transporter. This information was to be contained on a manifest (or form) that would accompany the waste through its travels. Originally, the EPA did not propose a nationally uniform system, but allowed states to develop their own manifests. In response to complaints (primarily from industry) concerning the differing requirements from state to state, the EPA developed a national manifest.[18]

The clear intention of the manifest and recordkeeping requirements was to inhibit the practice of midnight dumping, or illegal disposal, and to firmly fix responsibility for the ultimate disposition of wastes. For example, a transporter cannot accept hazardous wastes unless they are accompanied by a manifest, nor can a treatment facility accept wastes from transporters unless accompanied by that same manifest.[19] This system could be used to develop a data base on the amount and type of wastes being generated in this country, but there is little evidence that the manifest requirements are serving such a purpose. Moreover, since the overwhelming percentage of hazardous wastes never leaves the point of generation, only a small portion of these wastes falls under the purview of the system. Further, we know very little about the effect of the manifest system on inhibiting illegal disposal of wastes. The incentives for midnight dumping still exist under the manifest system, at least for those finding illegal disposal cheaper than contracting with a responsible waste transporter and disposal site.[20]

The heaviest burden under RCRA (1976) fell on those facilities engaged in the treatment, storage, and disposal of hazardous wastes. The act required the EPA to establish performance standards for treatment, storage, and disposal facilities that are "necessary to protect human health and the environment."[21] RCRA further required that these standards be applied to, but not limited to: recordkeeping, reporting, monitoring, and inspection; treatment, storage, or disposal methods, techniques, and practices; location, design, and construction of facilities; and a host of related matters. Amendments to RCRA adopted in 1980 allowed the EPA to distinguish between new and existing facilities in setting regulatory requirements.

The Environmental Protection Agency's response to the RCRA (1976) requirements has been characterized by fits and starts. Nine years after RCRA was first passed, the law had not been fully implemented. In fact, the first set of major regulations governing TSDFs was not promulgated until 1980. These rules set standards for operation on an interim basis until the final regulations were issued. Many of the requirements for facilities wishing to continue accepting wastes would already have been met by any reasonably well-run facility. For example, under the interim rules, TSDFs were to comply with the manifest requirements, maintain certain records on wastes handled over the life of the facility, and design and implement an inspection program. More economically burdensome were those elements of the interim rules designed to reduce potential long-term risks. These included the EPA-imposed post-closure and financial responsibility requirements, liability insurance requirements, and groundwater monitoring activities. Taken together, these elements of the interim rules were the EPA's first attempt to reduce the likelihood that existing sites would become abandoned dumps. However, the post-closure rules and financial responsibility requirements may have had unforeseen consequences, as we will see.

In July 1982, two years after promulgation of the interim standards, the EPA issued the final technical standards to be incorporated in permit applications for disposal facilities. These standards were intended to be the agency's ultimate interpretation of the law governing TSDFs. While it is not possible to outline every requirement of the 1982 regulations, some of the more important are:

- Incinerators: facilities that burn hazardous wastes must achieve a 99.99 percent reduction of principal organic hazardous constituents. (The regulations also set maximum emissions rates or minimum removal rates for several other types of constituents.)
- Landfills: land-based disposal facilities containing or receiving hazardous wastes must be equipped with liners, must have collection systems above the liner to trap wastes that may leak out, and must adhere to certain inspection, monitoring, and post-closure programs.
- Surface storage tanks: all tanks must have containment systems to minimize leaks or spills from the tanks. The actual system is to be designed on a case-by-case basis.

These requirements formed the basis for the issuance of final permits allowing the operator of a facility to keep it open. The intention of Congress was to have the states carry out the permit process after receiving approval of their hazardous waste management plans from the EPA. These plans could differ from the federal program as outlined in the implementating regulations, but would have to be, in the words of RCRA, "substantially equivalent."[22]

The final technical standards of 1982 are described by the EPA as performance standards, and they are much less specific than an earlier set of proposed rules. In regard to actual design requirements, they leave substantial discretion to the permit writers, and they provide an opportunity for tailoring the requirements to meet site-specific needs. Because so few TSDFs have final permits, it is impossible to know exactly how the process would have worked had it not been changed by the 1984 amendments. It remains unclear how national uniform standards—even those specifying only goals—could be sensitive to the wide variations in the type of sites and type of wastes produced and processed, as well as the varying levels of risk associated with the facilities.

The 1984 RCRA Amendments

In 1983 Congress began to reevaluate the hazardous waste program as it had evolved under the 1976 and 1980 amendments. Against a backdrop of very slow or nonexistent progress under the earlier legislation, increasing public concern, conflicts with the Superfund law of 1980 (see the next section), and upheaval at the EPA, it is little wonder that Congress passed amendments to RCRA in 1984 that can only be described as among the most detailed and restrictive environmental requirements ever legislated.[23] More so than in any previous environmental law, Congress took on the role of regulator itself and set out specific instructions, rather than general guidance, on the form the hazardous waste programs should take. It was a precipitous step away from delegation of powers to an administrative agency. No major section of the act went unchanged, and several new major programs were added.[24]

Congress had four goals in mind in the 1984 overhaul of RCRA. First, it wanted a program that was heavily biased against the land disposal of hazardous wastes—the disposal method most often associated by some with environmental risk. Second, it wanted the most far-reaching of the regulatory exemptions granted under the 1976 program revoked, and the universe of control expanded to include more sources and wastes. Third, it wanted the program to get under way with no further delays. And fourth, it wanted to address the problems that might arise if and when active waste disposal facilities discontinued operation. There was little sympathy given in these amendments to the lessons learned from the fourteen-year experience with air and water pollution control. Thus opportunities to reduce the costs of meeting environmental goals were ignored in the interests of quick and dramatic action.

As always, it is difficult to summarize succinctly such a complicated statute. Nevertheless, the 1984 amendments made three major changes worth discussing in some detail. To begin with, they closed what had come to be viewed as a loophole in RCRA by bringing under regulation an estimated 130,000 relatively small sources that generate between 100 and 1,000 kilograms of hazardous waste per month—sources exempted earlier by the EPA in the interest of manageability. However, Congress did recognize that regulation of these small sources might require a different approach than that applicable to large-scale waste producers. Thus the regulations for small sources required under the 1984 act are allowed to differ from those set for larger generators, but they still must be "sufficient to protect human health and the environment."[25] In the meantime, small-quantity generators must come into compliance with the EPA's uniform manifest requirements, and all of their wastes must be disposed of in a facility that is permitted to receive municipal or industrial solid waste.

Another major change introduced in 1984 concerned the disposal of certain hazardous wastes on land. In the eyes of some, the biggest shortcoming in the evolving hazardous waste program was a bias toward land disposal of wastes. With public concern mounting over the perceived risks of land disposal (such as the increasing number of abandoned disposal sites), and a report by the Office of Technology Assessment which argued that alternative disposal techniques were available, Congress decided to force the EPA to adopt a bias against the disposal of hazardous wastes in landfills.[26] Specifically, RCRA was amended to prohibit land disposal of hazardous wastes unless the EPA administrator determines that "the prohibition of one or more methods of land disposal of such wastes is not required in order to protect human health and the environment for as long as the waste remains hazardous."[27]

The basis for making such a determination under the 1984 amendments is confusing and complex. Essentially, the EPA must ban all untreated hazardous wastes from landfills unless it can be demonstrated that the landfilling method will meet certain restrictive conditions (among them the requirement that wastes will not be allowed to migrate from the landfill for as long as they remain hazardous). The shift from the usual regulatory presumption of "safe until proven hazardous" to the position of "no land disposal unless proven safe" is the most obvious manifestation of congressional desire to do away with land disposal. It would be difficult to prove that some kinds of wastes are unable to escape from land disposal sites. While an exemption in RCRA for pre-treated or innocuous wastes may offer some relief from the regulatory strictures, the burden of proof is still severe. The eventual impact of this requirement cannot yet be foreseen, but the overall standards of safety under the 1984 amendments are extremely demanding.

A third major change introduced by the 1984 amendments focused on a number of potential sources of hazardous waste pollution that were not recognized as such under the earlier RCRA program and were thus left unregulated. Among them were underground storage tanks used to hold petroleum, solvents, pesticides, and the gasoline at service stations. Evidence became available just before passage of the 1984 amendments that as many as 100,000 such tanks, containing a wide range of potentially hazardous materials, may have been leaking their contents into the surrounding environment. Fueled by the fear that many of these tanks were or might become serious environmental hazards, Congress responded by bringing all underground storage tanks into the RCRA system. The EPA is thus required to establish standards for detecting leaks, to remedy the problems that lead to leaks, to determine financial responsibility for tanks that may be taken out of use, and to issue rules regarding the design and construction of any new underground tanks. The EPA has estimated that about 1.4 million tanks will be subject to these rules. Like most of the 1984 RCRA amendments, this one provides no explicit directives that would allow for consideration of variations in risks or costs across the regulated community.

Congress was not content in 1984 to simply list regulatory requirements; it also wanted to force action on the EPA's part. Dissatisfied with the slow pace of earlier EPA activity under RCRA, Congress laid out a specific timetable for the agency to accomplish the tasks delegated to it under the 1984 and earlier amendments. For instance, the EPA was required to identify acceptable disposal techniques for one-third of all high-volume, highly hazardous wastes by August 1988; do the same for the other two-thirds of those wastes by May 1990; and address disposal of all other wastes by April 1991. Action on certain wastes was put on even faster schedules; for example, the EPA was to determine whether to allow land disposal of dioxins and solvents by November 1986.

In addition to establishing these deadlines, the 1984 amendments spelled out specific regulations that would automatically take effect unless the EPA devised its own regulations in their place by certain other deadlines. For example, unless EPA acted affirmatively to establish conditions under which certain wastes could be disposed of in landfills, these wastes would automatically be banned permanently from such sites.

Congress added numerous other provisions and deadlines to RCRA in 1984. By closing loopholes, bringing more substances and entities under the regulatory framework, and laying out regulatory plans in great detail, Congress took far-reaching actions in an attempt to force the internalization of the costs of improper disposal. Yet this was done with very little idea of what those costs might be; moreover, Congress provided little flexibility for the EPA to ease regulations where costs might be large relative to anticipated benefits.

The long-term response by the regulated community to the 1984 RCRA amendments is difficult to predict at this early stage. There is some evidence that the large, better-financed firms are beginning to shift production processes and inputs to reduce the amounts of waste produced.[28] This is clearly what Congress had in mind. On the other hand, the current RCRA system embodies few of the broad economic concerns discussed above. Flexibility in the design and implementation of regulatory strictures is not a hallmark of the RCRA program. Rather, the RCRA approach to hazardous waste management calls for detailed specification and regulation of virtually every aspect of disposal facilities and disposal options. RCRA is not going to lead to much balancing of costs and benefits. Moreover, recent attempts by the EPA to introduce some balancing of costs and benefits in the landfill prohibition regulations have met stiff opposition from environmentalists and many lawmakers. The regulatory decisions have already been made by Congress. Yet there is little evidence that the data to judge the economic merits of the RCRA structure were available to Congress. The incentives embedded in the act suggest that a number of varying responses, some not so constructive, should be anticipated (see the section on Hazardous Waste Management, below). From an economic perspective, the outlook for hazardous waste disposal under RCRA is not encouraging. A recent study by a bipartisan legislative organization concluded that regulatory uncertainty, lack of data, and inadequate disposal capacity are likely to limit the ability of RCRA to achieve its goals.[29]

The Comprehensive Environmental Response, Compensation, and Liability Act

In 1978, two years after the passage of the initial RCRA, the nation began to realize that the generation and disposal of hazardous wastes was not something new. It was then, in a residential area of Niagara Falls, New York, called Love Canal, that foul-smelling chemicals and other substances were found seeping into the basements of homes. Rocks struck against the sidewalks would send off colorful sparks, and the drinking water tasted and smelled peculiar. Upon investigation it was determined that these homes and a nearby school had knowingly been built on donated land above an industrial waste disposal site that had long been closed. As other sites became known to pose environmental risks, the nation also became aware that the by-products of rapid economic growth in the 1950s and 1960s must have been disposed of somewhere. The problem of old hazardous waste disposal sites was not confined to an isolated environmental event. More and more sites were being uncovered where wastes had been disposed of years before.[30]

At first it was assumed that such remnants of prior disposal practices would fall under the regulatory strictures of RCRA, but it later became obvious that RCRA was inadequate to deal with abandoned sites. For one thing, the abandoned sites were not going to disappear as a result of regulations designed to curb future problems, yet the latter were the main focus of RCRA. For another, it was not always clear who was responsible for having disposed of wastes at an abandoned site. Many sites had been operated outside the conventional disposal system and their owners were hardly likely to come forward. Even when a number of the parties that had used a disposal site could be identified, tens or hundreds of other users of the same site might go unidentified. Unless very careful records of the receipt of wastes had been kept—which was unlikely at all but the best-run disposal sites—identifying who had dumped what was nearly always impossible. Finally, the existing legal remedies through common law nuisance or negligence actions were judged by some to be too time-consuming and cumbersome to handle what were perceived as the immediate and life-threatening risks of abandoned hazardous waste sites. A new and novel program appeared to be required.

The response by Congress was passage of the Comprehensive Environmental Response, Liability, and Compensation Act of 1980 (or CERCLA, better known as Superfund).[31] There is perhaps no more telling evidence of the supercharged political atmosphere that begat Superfund than that it was passed by a lame-duck Congress during the transition from a Democratic to a Republican administration. Whereas other environmental programs evolved over many years, Superfund emerged much more rapidly, a monument to public concern that had no precedent in environmental law. The 1980 act was amended in 1986 by the Superfund Amendments and Reauthorization Act, or SARA.[32] Much of the following discussion is focused on the original Superfund, the basic thrust of which was unaltered by SARA, but the key changes introduced by the 1986 amendments will be highlighted.

The Superfund law is unique in at least two respects. First, it is one of the few environmental statutes that attempts to address past environmental degradation rather than prevent future pollution. In other words, its focus is ameliorative rather than preventive. Second, Superfund places the Environmental Protection Agency in a unique position: the agency is on the one hand cast in its classic role of regulator, stipulating the conditions and requirements under which cleanup of hazardous waste sites will take place; on the other hand, the EPA is called on to act as a hazardous waste engineering firm, actually conducting site cleanups subject to those regulations. The two roles are not necessarily in conflict. However, the statute does require that the EPA wear two hats and set the rules under which it must itself operate.

Superfund has several major provisions, the most important of which gives the EPA authority to identify parties responsible for currently inactive or abandoned wastes sites and force clean-up actions, or to clean up those sites itself and find (or sue) responsible parties later. Superfund also sets up a process by which sites can be identified and ranked by priority for clean-up action. The EPA can undertake emergency action or institute short-run or long-run remedies designed to mitigate any risks to health and the environment. In general, the law prior to the 1986 SARA amendments was unusually generous in the flexibility it allowed in determining what to clean up at a particular site, how and when to do it, and what constituted an adequate cleanup. The steps in the Superfund process are illustrated in figure 5–1, which also shows the average time taken between certain of these steps. It is a process that can take twelve or more years to complete.

The Superfund act also required the EPA to prepare a plan (now called the National Contingency Plan, or NCP) that stipulates the processes for selecting sites for response, conducting investigatory analyses, selecting appropriate remedies, determining appropriate levels of cleanup, determining who will pay for remedies, and determining procedures for ensuring that a site will not pose risks in the future.

The EPA and the states nominate sites to be included on a list of those determined to be in greatest need of some form of response—the National Priority List. NPL sites are subject to the clean-up requirements set out in the National Contingency Plan. To select among nominated sites, the EPA developed a hazard-ranking system which assigns numerical weights to certain characteristics of the sites, such as the toxicity of the chemicals contained at a site, and the possibility of human exposure. As of

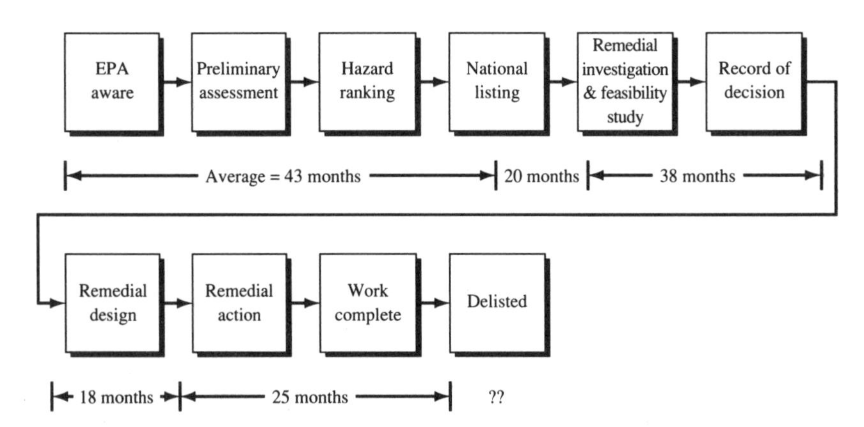

Figure 5–1. Principal steps in the Superfund process. *Source:* Jan Paul Acton, *Understanding Superfund* (Santa Monica, Calif., The RAND Corporation, 1989) p. 16, fig. 3–3.

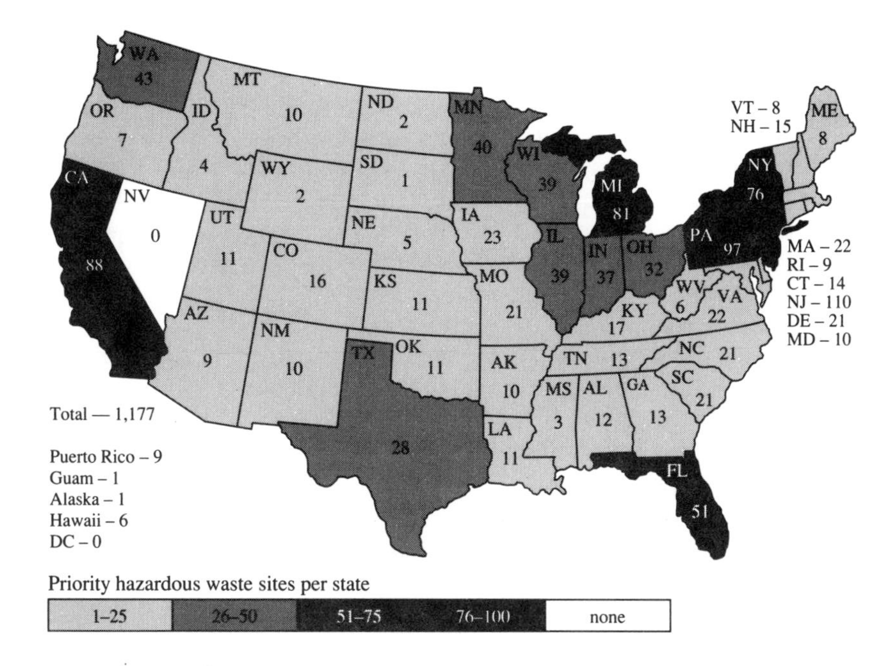

Figure 5–2. Location of 1,177 EPA Priority Hazardous Waste Sites, as of June 1988. Numbers indicate both actual and proposed National Priority List sites.

mid-1988, there were 1,177 actual and proposed hazardous waste sites on the National Priority List, including 32 federal facilities. About 50 percent of the sites were located in seven populous states—New York, Michigan, New Jersey, Pennsylvania, Minnesota, California, and Florida. As figure 5–2 illustrates, all parts of the country (and all but one state) are represented on the list.

The sites currently on the EPA's list are not necessarily the worst in terms of actual risks to health and the environment. The actual exposures of individuals are not considered in ranking sites. Moreover, opportunities for pork-barrel allocations are not absent from the ranking process; the process is heavily biased toward potential health risks rather than environmental risks; and the National Priority List contains only sites covered under the current statute—it does not include, for example, sites where agricultural pesticides alone are involved.

Once on the National Priority List, the process by which a site is selected for remedy and a specific clean-up option is somewhat less structured. Selection appears to be a function of information available on the site characteristics, the degree of risk involved, the potential for private rather than public action, the cost-effectiveness of the action, the opportunity cost of the action, and political pressure.[33] There is perhaps no typical Superfund action, but for illustrative purposes table 5–7 depicts the main steps of the clean-up action at a hazardous waste site in Utah.

Since the passage of Superfund, the pace of remedial actions under the act has been slower than expected. Of the 30,000 sites requiring preliminary investigation for ranking, about 90 percent have been investigated, but the EPA has stated that of the sites requiring cleanup, only 48 have been completely cleaned up.[34] A more complete breakdown of clean-up activity at Superfund sites is shown in table 5–8. Most activity under Superfund has involved site studies (remedial investigation and feasibility studies) and waste removal. In the latter, existing wastes are exhumed and taken to facilities having permits. Moreover, even these estimates are suspect by some who feel that the EPA's definitions of a completed action or a cleanup do not reflect the potential for or recurring problems at sites and the need for possible future actions.[35]

How clean a site must be in order to be considered remedied is a topic that has vexed the EPA over the last nine years. The first regulations under the Superfund act required that all remedial actions result in levels of environmental improvement comparable to the standards established under other existing federal environmental laws. Thus the cleanup of contaminated groundwater was not considered complete unless the pollutant concentrations in the treated groundwater were less than the maximum contamination levels set under the Safe Drinking Water Act. Some felt, however, that federal clean-up standards should be designed to reduce health and environmental risks to zero. Others urged the EPA to consider matching the degree of risk posed by a site with the costs of reducing that risk—which raised the possibility that some risks would be cleaned up more than others. The Superfund Amendments and Reauthorization Act resolved some of this debate by requiring that site cleanup be subject to federal standards under other statutes as well as all state standards. Cleanups that involve permanent treatment, as opposed to temporary removal, are preferred. The degree to which the selection of a clean-up option can be guided by cost considerations is still a matter of debate and will not be resolved until the EPA revises its guidelines. Cost-consideration choices range from evaluating the cost-effectiveness of alternative clean-up technologies within a given treatment level to not allowing costs to be evaluated at all.

The current statutory approach to the question, How clean is clean? raises directly and indirectly several issues that have characterized the past and present debate. First, what happens when a site contains contaminants for which no current standard is in place? How far should cleanup go?

Table 5–7. Key Steps in Remedial Action at Rose Park Sludge Pit, Utah

Time period	Action
1920–1957	Acid sludge from petroleum refining operations dumped into pit on a 5–6 acre site.
1957	Citizen complaints prompt Salt Lake City to purchase site; disposal discontinued. City removes 100 to 200 truckloads of sludge; remaining sludge covered with soil cap. Site included in land used for development of golf course.
1976	Expansion of park facilities leads to breakage of cap and numerous extrusions of sludge. Workers complain of respiratory and eye irritations. Heightened community fears of potential health hazard lead city to enclose pit with a 4-foot fence.
1977	City requests Utah Geological and Mineral Survey to investigate groundwater contamination. UGMS concludes pit "seems" to be contributing to contamination of groundwater.
1979	EPA Region VIII Technical Assistance Panel studies extent of groundwater contamination. Analysis shows presence of various priority pollutants; sludge boundries extend beyond fenced area. Fence extended to include contaminated areas.
1979–1981	EPA conducts a series of field investigations. Community concern grows, focusing on physical contact with sludge extrusions, the release of acid vapors from site, groundwater contamination, and potential decline in area property values.
January 1981	Mayor's office contacts EPA to discuss application of Superfund to Rose Park Sludge Pit. Site placed on National Priority List as Utah's number 1 site.
March 1981	Amoco Oil Company (one-time owner of site) volunteers to take action at the site. Amoco initiates a comprehensive field investigation, signs an Intergovernmental/Corporation Agreement (ICCA), and agrees to implement remedial action.
January 1983	Circumferential slurry wall and containment cap to prevent migration of sludge from pit completed.

Second, does the current approach take into consideration possible synergistic relationships among the many chemicals found at hazardous waste sites? Third, the economic approach to the how-clean-is-clean question would allow some balancing between the levels of risk at any given site and the costs of bringing those levels down. Should this be permitted? The tendency under the current approach is to bring all sites up to a nationally uniform level of cleanliness. The question remains whether funds, from public or private sources, are best spent by cleaning up all sites to a predetermined level or by differentiating among the risks associated with various sites and concentrating resources on the worst risks first.

Finally, do the current guidelines for cleanup provide an incentive or opportunity for developing and using new technologies that, while untested or of uncertain effectiveness, may result in much more cost-effective responses? The record to date on this question is unclear, but it

Table 5–8. Numbers of Sites at Major Stages of the Superfund Process, Cumulative through March 1989

Status of sites	Number of sites
Sites in the EPA information system	30,844
Preliminary assessments completed	28,101
Site inspections completed	9,902
Sites with no further action planned	12,416
National Priority List	
Final	890
Proposed	273
Total	1,163
Removal actions	
NPL	274
Non-NPL	1,073
Total	1,347
Remedial investigation or feasibility study, cumulative starts	845
Remedial design, cumulative starts	300
Remedial action, cumulative starts	204
Site work completed	41
Delisted from NPL	26

Source: Environmental Protection Agency, "Superfund Progress Report" (March 1989), as reprinted in Jan Paul Acton, *Understanding Superfund* (The RAND Corporation, 1989) p. 61, appendix A.

would not be surprising to see a certain bias toward expensive but "known" clean-up alternatives.

In addition to its clean-up provisions, Superfund deals with an explicit economic issue—the means of financing cleanups of abandoned sites. Congress assumed that most Superfund sites would be cleaned up by the use of private funds once the EPA identified the responsible parties. However, Congress also recognized that for some sites it would never be possible to locate the responsible parties, and that for others, clean-up actions would be required long before the courts had determined the relative burdens of the contributors. The solution was the establishment of a trust fund on which the EPA could draw to finance public cleanups. Originally the clean-up fund was set at $1.6 billion and was to be financed by three principal sources: private funds from the parties held to be responsible for the dump site; public funds from general tax revenues; and monies raised from a special tax, created in the act, on petroleum and chemical feedstocks. The fund would be replenished as the EPA identified and forced responsible parties to pick up their share of the clean-up tab.

The feedstock tax is a varying, per-unit levy on a wide range of primary inputs to the production of chemicals and petroleum derivatives. It was apparently never considered to be an economic incentive for reducing the generation of hazardous wastes. Rather, Congress searched for a reliable and relatively uncontroversial source of dollars to finance the fund. While there was some notion that the petrochemical industries were primarily responsible for the generation of the kinds of wastes now contained in Superfund sites, the fact that these industries were financially sound was also an important factor. The debate at the time focused mostly on the effect of the proposed taxes on the industries' earnings, and not the equity of requiring them to foot a large part of the bill for past disposal problems.

SARA expanded the fund to $8.5 billion over five years and broadened, to a certain degree, the sources of income. Both the petroleum and chemical feedstock taxes as well as general revenues are still important components, but SARA also included an "environmental tax"—an assessment on every domestic corporation of 0.12 percent of the corporation's minimum taxable income over $2 million. The statute assumes that around $2.5 billion will be available for the Superfund from this tax.

Perhaps more important, ultimately, than the direct clean-up provisions of the Superfund act is its impact on the legal environment in which clean-up decisions are made. This is particularly true in regard to the assessment of liability for abandoned disposal sites. The act imposes what is known as strict, joint, and several liability on responsible parties. Simply put, this means that the EPA can hold one party whose wastes were disposed of at a particular site responsible for *all* the costs associated with cleaning up the site—regardless of the share of total waste disposed of at the site by the identified firm or the level of care given by the firm to the disposal activity. The burden of proving that something "wrong" was done and of recovering damages from the other responsible parties is shifted to the firm initially held responsible, thus easing the financing and legal obstacles to the government of initiating a cleanup at a site.

Besides creating a liability scheme to cover cleanup of existing hazardous waste sites, Superfund also establishes a liability strategy for dealing with damages to the environment that may occur before and after a site is discovered and cleaned up. Under these rules, local, state, or federal governments may seek dollar compensation from responsible private parties for natural resources that are injured or destroyed by spills and releases of hazardous wastes. Although these natural resource damage rules and requirements have only recently been formalized and as yet have seen little use, they represent a significant expansion of the potential liability facing individuals and firms that deal with hazardous wastes.[36]

The importance of these liability provisions cannot be overstated. Since many Superfund sites have been used by several, and in some cases hundreds, of responsible parties, the EPA would expend much of the Superfund monies trying to build cases against each one. In the recently settled Chem-Dyne case, for example, where more than 200 parties were potentially responsible, the cost of the possible legal proceedings needed to establish the liability of each party was estimated to exceed the actual cost of cleanup. Under the liability provisions of the Superfund act, the EPA or the states need only identify the most significant or obvious contributors and then construct a legal case as if those contributors were responsible for the entire problem.

It should be noted that one does not need to have owned or participated in the use of a Superfund site to be caught under the law's liability provisions. In a number of situations, institutions that lend money to firms that may have operated or used a hazardous waste site can also be held accountable. Here the requirements apply to federal lending agencies as well as private ones. The potential liability of the government as a lender could vastly increase the number of sites for which federal agencies have financial responsibility for cleanup.

The liability and enforcement provisions of Superfund are intended to provide incentives for private actions to supplement the fund-financed cleanups. It was always hoped that there would ultimately be more private than public remedial actions. The number of settlements involving private cleanups on Superfund sites accelerated from 13 in 1981 to 565 by the end of 1988, and entailed costs of around $1.2 billion.[37] The number of future settlements, however, will be affected by new settlement provisions embedded in SARA, one of which allows relatively small contributors to a Superfund site to settle early for their share of the clean-up costs.

HAZARDOUS WASTE MANAGEMENT: CRITICAL ISSUES

The Resource Conservation and Recovery Act and the Comprehensive Environmental Response, Compensation, and Liability Act together establish a state-federal partnership for coping with past and present disposal activities. The acts also intend a private role in this partnership, although the Superfund act is aimed most directly at creating a bureaucratic response to abandoned toxic waste dumps. The two acts offer the potential for a life-cycle approach to hazardous waste generation and disposal. Yet, as our preceding discussion suggests, many questions remain unanswered and the ability of the two programs to achieve their stated goals is quite uncertain. Not surprisingly, such complex programs have a wide array of hidden (and some not so hidden) incentives and effects on the regulated community and on public health and welfare. An economic-based approach to controlling hazardous waste risks would try to take advantage of economic incentives to reduce risks while balancing the costs and benefits of doing so. In this section we examine several features of the current system that are likely to result in a deviation from this approach.

The Pace and Cost of the Programs

We have already commented on the slow pace of controlling hazardous wastes under RCRA (1976) and on the effect this had on the 1984 amendments. Nine years' experience with the Superfund act is no more encouraging. Since 1980, the EPA has completed remedial activities at only 48 sites out of a universe of 1,177 proposed or final National Priority List sites.[38] Depending on how the scope of Superfund is eventually defined (does it cover underground storage tanks or municipal sanitary landfills?), the total universe may be greater than 100,000 sites. Although in launching any new program of Superfund's scope one would expect delays and a slow start, few observers appear satisfied with the progress under the act to date. Even though the pace appears to be improving under the deadlines imposed by SARA, real prospects for speeding up are not encouraging. The EPA expects the average length of time required to complete a hazardous waste site cleanup to grow from 58 months to 67 months under the provisions of the new law.[39]

In reauthorizing Superfund in 1986, Congress responded to the slow Superfund start with a statute that included a timetable for conducting a specified number of site studies and site cleanups. While this response may be understandable, it might also be premature until more is known about the risks posed by individual sites and the best means for cleaning up those sites. There is no point in moving wastes from one place to another if the problem is only shifted around. The potential opportunity costs of going too far too soon could be significant. In this light, the slow start of Superfund cleanups may have made some sense to the extent that the early years involved a substantial learning experience.

It is instructive to examine one study in which the costs of cleaning up three hazardous waste sites were estimated and compared to the estimated costs of bringing the sites into compliance with the 1976 RCRA requirements (thus they would not have posed a Superfund problem).[40] At two of the sites, the costs of prevention (that is, of complying with RCRA) were estimated to be greater than the clean-up costs. Besides calling into question the standard bromide that an ounce of prevention is worth a pound of cure, the study illustrates the wide range in costs and risks associated with different actions at hazardous waste sites. It highlights the need to proceed with caution in implementing programs where the economic stakes may be quite high. Neither RCRA or Superfund provides much opportunity, however, to build an information base that would assist in regulatory design.

The costs of our current hazardous waste programs are large and growing. For 1983 alone, industrial expenditures to comply with the 1976 RCRA regulations were estimated by the Congressional Budget Office to be almost $6 billion.[41] The 1984 RCRA amendments are guaranteed to substantially increase compliance costs. While there are no firm estimates of the costs of complying with the final regulatory requirements of the 1984 amendments, the EPA has estimated that the compliance costs associated with selected provision of RCRA to be around $2.2 billion a year (see table 5–9).[42] Restrictions on landfill disposal alone may have an annual cost of $1.6 billion. The Congressional Budget Office has estimated that the additional annual costs associated with the 1984 amendments to RCRA may be in the range of $3 to $7 billion by 1990.[43] These estimates should be taken with some caution. For example, the 1984 RCRA corrective action program requires clean-up, monitoring, and financial responsibility compliance for a broad range of waste-receiving facilities and has been called a mini-Superfund. Some argue that the costs for this program may dwarf current RCRA compliance costs, although the EPA estimates this cost at around $.5 billion per year.

Table 5–9. Estimated Additional Costs of the 1984 Amendments to RCRA (EPA estimates of the most likely cost)

Provision	Annualized additional cost (millions of dollars)
Liquids in landfills	8
Minimum technological requirements	63
Corrective action	476
Dust suppression	26
Small-quantity generators	48
Burning and blending of oil; used oil	21
Exposure assessments	a
Delisting procedures	a
Hazardous waste exports	a
Waste minimization	a
Financial responsibility	10
Restrictions on land disposal for banned waste	1,600
Total	2,252

Note: Some potentially major costs of the 1984 RCRA amendments have not been estimated yet because the regulatory option has not been chosen. Two examples are the final standards for underground storage tanks and the subtitle D program.

Source: J. E. McCarthy and M. E. Anthony Reisch, *Hazardous Waste Fact Book*, 87-56 ENR (Congressional Research Service, 1987). Estimates were derived from the *Federal Register*, January 11, 1985, July 15, 1985, March 24, 1986, and November 7, 1986. Data were supplemented and in some cases revised by information provided by the Economic Analysis Staff of EPA's Office of Solid Waste.

[a] Less than $500,000.

The RCRA amendments are also having an impact on the EPA's own resources. For instance, the dollar expenditures required of EPA to implement RCRA increased from $184 million to $265 million between 1985 and 1989.[44] There is little doubt that the public and private costs of compliance with RCRA are now beginning to mount. While they are still below those arising from the Clean Air and Clean Water acts, they are now of the same order of magnitude and are increasing at a much faster rate.

As Superfund expands in scope, the costs associated with it will also grow. Any specific estimate depends crucially on assumptions concerning the number of sites covered, the cleanup required at each site, and a host of other factors. Nevertheless, even using the EPA's conservative estimate of $7 million per site addressed and assuming that 2,000 sites are eventually cleaned up (using federal funds), charges against the fund may come to $14 billion. More recently it was estimated that, as a result of the 1986 amendments to Superfund, the average cost of a cleanup will rise to $30 million. If this is so, cleanup of 2,000 sites would come to $60 billion.

The estimates above do not include the costs incurred by private parties in undertaking remedial or emergency actions, or the legal costs associated with determining liability and the cost-sharing of private cleanups. Such costs vary widely. Of 23 private and public remedial actions studied in one report, the costs of cleanup ranged from $23,000 to $10 million, with an average cost of $1.5 million.[45] Many of these sites, however, were cleaned up in the early 1980s before the formal Superfund requirements went into effect. The General Accounting Office has estimated that if 4,170 sites require cleanup under Superfund and SARA (a high estimate derived by the EPA), total federal, state, and private costs could eventually total $68.0 billion (in 1984 dollars, discounted at 5 percent).[46] If we were to add an additional 25 percent in costs as a conservative estimate of litigation expenses, the total cost of cleaning up past disposal problems could be in the range of $80 billion. This estimate will be on the high side, however, if some identified sites do not get the most expensive cleanup or if technological advance lowers the cost of cleanup over time. On the other hand, the estimate does not include EPA or state administrative expenditures or those of the private sector, or the costs of damages to natural resources.

Moreover, the costs associated with dealing with the hazardous waste problems that exist at federal facilities should be added to this list. While there is still tremendous uncertainty concerning the eventual scope of such problems at federal facilities (and cost estimates at this time do not break out RCRA versus Superfund expenditures), preliminary estimates of costs to eight agencies of complying with RCRA and Superfund over the period 1990 to 1994 are in the neighborhood of $10 billion, as shown in table 5–10. The Department of Energy alone has estimated its eventual costs of complying with RCRA and Superfund at as much as $91 billion. These expenditures will in large part be devoted to cleaning up industrial and low-level radioactive contamination at nuclear weapons production facilities.[47]

Balancing Benefits and Costs

The increasing costs associated with the RCRA and Superfund programs suggests the real need to carefully balance where and how money is spent against the level and type of risk reductions obtained. This need is even more pronounced when one considers the wide range of health and environmental risks associated with the different sites, different substances, and different control measures. Yet RCRA—and to a lesser extent Superfund—is a statute primarily driven by health risks, and the mandates of these laws explicitly rule out such balancing in many key

Table 5–10. Federal Agency Estimates of Costs of Compliance with RCRA and Superfund, Fiscal Years 1991 through 1994

Agency	Costs (millions of dollars)
Department of Defense	2,000
Department of Energy	7,670
Department of Agriculture	55
Department of the Interior	22
Department of Justice	13
National Aeronautics and Space Administration	120–185
Postal Service	160
Department of Transportation	300
Total	10,340–10,410

Source: Congressional Budget Office, *Hazardous Waste Liabilities at Federal Facilities* (Washington, D.C., forthcoming).

requirements. To a large degree, RCRA treats all wastes and all waste facilities as posing the same relatively high risks. Variations in the health or environment risks posed by regulated substances and activities are presumed to be minimal, or at least of little economic significance. The potential for misusing scarce resources is therefore tremendous. For example, in a preliminary study of the EPA's options for controlling incineration of hazardous wastes, the agency estimated that waste toxicities at incinerators can span eight orders of magnitude, and that the potential exposed population could range from 4,000 to 1.5 million. Yet all incinerators are required to meet the same best-available-technology standard.[48]

Variations in risk lead to variations in the benefits of regulatory control, yet we have little solid information or evidence on what the magnitude of these benefits might be. The lack of careful risk assessments associated with the cleanup of old or new hazardous waste sites has already been noted. Recent research sponsored by the EPA on the economic benefits of hazardous waste control has pointed out how difficult it is to make such assessments. Nevertheless, one careful study attempted to identify the reduction in lifetime cancer risk associated with the cleanup of an abandoned hazardous waste site in New Hampshire.[49] It found that as a result of a $25-million-dollar cleanup, the reduction in risk amounted to one life saved every 25 years. This cost-per-life-saved compares unfavorably with the costs of other environmental regulatory programs aimed at reducing cancer risks. Portney has recently estimated, for instance, that

$25 million spent on radon mitigation would reduce lifetime cancer incidence by 10,000.[50] However, cleaning up a hazardous waste site would result in other benefits that would not arise from a radon mitigation program. Efforts at benefit estimation also have included analyses of property value differentials related to hazardous waste sites, and valuations of hazardous waste risk reductions through survey techniques. One recent study of three hazardous waste sites suggests that the benefits from cleanup of each would be $3.6 million, $7.0 million, and $17.4 million at the three sites respectively, based on estimates of the increases in property values expected to result from the cleanups.[51]

There is little doubt that the state-of-the-art of risk assessment and benefit estimation will improve over the next several years. However, the lack of solid data on risks will continue to limit our understanding of the benefits of hazardous waste regulations.

An improved understanding of the benefit side of hazardous waste regulation is not simply a matter of having better data. Economic benefits are a function of public perception concerning environmental health risks. Currently, many members of the public perceive the risks associated with hazardous waste disposal sites to be extremely high, and it is these perceptions that are driving the current system. We know, however, that the relationship between perceived risks and actual risks is quite uncertain.[52] For example, a recent study of property values around an industrial landfill in California showed large property value impacts associated with the perceived risks of the landfill, while several scientific studies of the site were unable to demonstrate any potential risk of harm. The simple announcement that the site would be closed resulted in a substantial increase in property values even though actual health risks would be unchanged (or at the most very small). The point of this example is not that individuals' reactions to perceived risks are wrong, but rather that it may not always be necessary to expend engineering or regulatory control resources to achieve economic benefits. Public education campaigns concerning hazardous waste risks might narrow the difference between perceived and actual risks and reduce the need to treat all sites as human health hazards.

We do not know, at this point, whether a reasonable balance exists between the costs and benefits of our domestic hazardous waste management strategy. The available information, although limited, suggests that they are not in balance. The combined costs to industry and government of Superfund and the RCRA may soon grow to rival the costs associated with the other major environmental programs. While it is too early to predict the impact on industry of such costs, it is likely to be significant, particularly in those industries most susceptible to international competition. Benefits, on the other hand, are uncertain. They may be significant if RCRA and Superfund are effective in preventing serious health and ecological problems. But the fragmentary evidence to date suggests that with respect to health, at least, the risks are not great. The hazardous waste management program is not the first environmental protection strategy that may involve serious economic inefficiencies, but it may eventually go further in that direction than any other.

Liability and Insurance

Like other environmental statutes, RCRA and Superfund rely as little as possible on economic incentives to accomplish their goals. Instead, they contain a familiar mixture of command-and-control strategies, as well as health- and technology-based standards. Relatively ignored in the regulatory structure are important market or legal mechanisms for inducing appropriate behavior. Two mechanisms—liability under tort actions and insurance requirements—are particularly useful in providing incentives for cautious handling of hazardous wastes. The two are best treated together.

Negligence actions under state or federal law (tort law) provide one potential avenue for victims of certain activities to be compensated for their injuries, as mentioned briefly in chapter 2. In the process of determining liability for certain damages, court rulings and awards serve to allocate the benefits and costs of activities. Under certain conditions, court actions can result in an allocation of resources that replicates that achieved by freely operating markets. A host of studies have concluded that these conditions for achieving a proper allocation of risks and benefits may not exist for health risks involving hazardous wastes and other toxic substances.[53] In such instances, legislators may sometimes attempt to compensate for these conditions and reduce the plaintiff's burden of proof by applying strict liability rules or adopting statutory changes that lower the burdens of establishing a case. Through the assignment of strict, joint and several liability, Superfund has made it somewhat easier to overcome judicial barriers to establishing liability for a cleanup.

Insurance can play a dual role by providing compensation to injured parties and imposing costs, in the form of higher premiums, on firms that operate unsafely or engage in dangerous activities. The latter are imposed through what is known as the experience rating of insured parties. If insurance markets are operating properly, high-risk generators of hazardous wastes would pay higher premiums for liability coverage than low-risk generators or disposers. An economic incentive is thereby created to minimize the sum of disposal and insurance costs. As a result, a firm would select a disposal method and a level of liability coverage that would begin to balance marginal costs and benefits.[54]

Under RCRA, all land-disposal TSDFs having current permits, most of which were operating on interim status, were required to have a certain amount of liability insurance by November 1985. At first glance, this requirement may have appeared wise, an effort to meld regulatory and economic incentives into one program. However, the insurance market has not kept pace with regulatory requirements. Faced with great uncertainty concerning their ultimate liability, insurance carriers have been unwilling to write policies. In combination with the RCRA monitoring requirements, the difficulty of obtaining insurance resulted in the closure of nearly 70 percent of the 1,600 land disposal facilities doing business before November 1985.[55] The prospects for obtaining insurance coverage against Superfund cost recovery or clean-up actions is no better. At this time, the insurance market is saying that no premium is high enough to cover the possible costs of liability under RCRA and Superfund.

On the surface, the insurance and liability rules in RCRA and Superfund would appear to satisfy, in part, the call for increased use of economic incentives in hazardous waste management. Nevertheless, the extent to which insurance and liability incentives achieve economically appropriate responses depends in large part on the laws and regulations underlying the incentives. We have argued here that both of these may go too far, too fast. The net result may be unexpected outcomes and unanticipated costs.

For example, the absence of insurance markets increases the possibility of a significant shortfall in disposal capacity for hazardous wastes. Although alternative treatment technologies are looming on the horizon and some are actually operational, research and development of non-land-based disposal options is lagging. Further, with the possible exception of deep-well injection, most other disposal methods currently available—such as ocean- or land-based incineration—are themselves subject to stringent regulatory requirements and may not be available where and when required. In fact, in early 1988 the most prominent firm involved in at-sea incineration of hazardous wastes dropped its bid for a permit to demonstrate the technology for use in the United States.

Other options open to hazardous waste generators in the absence of insurance are to change production processes to reduce the amount of waste generated, or to circumvent the legal system of disposal (by engaging in midnight dumping). Generators of hazardous wastes are already looking for ways to reduce their costs of disposal by producing fewer wastes. However, in light of slow enforcement under RCRA, and consequently the small chance of detection, midnight dumping may unfortunately appear to be a cheap solution for firms that have little regard for the law or the environment.

Another possible unanticipated effect of the liability and insurance rules under Superfund may be an incentive for firms to avoid disclosing information about sites where they previously disposed of hazardous wastes. Why should a firm step forward to admit contributing to a site when, under joint and several liability, it might get stuck with all of the costs of the cleanup? There may also be a disincentive to enter into agreements to clean up the sites. To the extent that Superfund is intended to identify and remedy past disposal problems as cost-effectively as possible, private assistance is crucial. Yet the existing incentives would appear to induce firms to take as long as possible to make information available concerning sites and to delay participating in a cleanup. With strict liability in force and an absence of adequate insurance to cover clean-up costs and other claims against a company, delaying tactics may be in a firm's best interest even in the face of possible civil and criminal sanctions. As costly as it might be, litigation might well be the preferred option of a firm when the expected value of entering into litigation is a reduced share of clean-up costs and a longer timetable for remedies.

The net effect of ever-expanding litigation is quite costly. There is evidence that litigation costs for some Superfund recovery actions are exceeding the costs of cleanup. At least one study has put the cost of litigation under Superfund at 55 percent of actual clean-up costs.[56] Besides slowing the pace of cleanups, litigation expenses can inhibit government recovery of clean-up costs. Between 1986 and the end of 1988, the EPA recovered only $166 million from private parties, or roughly 7 percent of the $2.4 billion spent on Superfund cleanups through that same period.[57] If, on the other hand, firms were encouraged through more flexible liability requirements or financial incentives to forgo litigation or to come forward voluntarily with information, these litigation costs might be substantially reduced.

Some forces are at work that may eventually ease the liability and insurance problems. SARA allows for lower public-to-private settlement payment ratios, thus encouraging, in theory, more voluntary participation by private parties. As improved information concerning the risks at hazardous waste sites becomes available, a more certain basis for determining appropriate premiums can be established. In addition, it would not be surprising to see firms seeking insurance banding together to form their own insurance companies; some large firms might be able to self-insure. The immediate problem of meeting the 1984 RCRA insurance requirements could be lessened through a loosening of the requirements, in combination with the changes in liability provisions.[58] In the long run, however, more accurate information on risks will be required.

Who Foots the Bill?

In the case of RCRA regulations, those individuals who have benefited in the past from the cheap disposal of hazardous wastes will begin to pay higher costs in the future as a result of regulation. This includes consumers like ourselves who have paid less for the products that give rise to wastes. Regardless of any balancing of benefits and costs, such internalization of costs makes good economic sense.

The situation is more complicated in the case of Superfund, which addresses disposal activities that took place in the past. The beneficiaries of economically cheap but environmentally unsound hazardous waste disposal practices in the past were those users of the associated products that generated the wastes—again, consumers such as ourselves. Yet the original funding scheme for Superfund imposed the bulk of the clean-up costs on consumers of products that use the taxed feedstocks as inputs (since the tax fell on the energy and chemical companies). These consumers may or may not have benefited from the careless disposal practices of the past. The products whose prices increased because of the feedstock taxes may or may not have been contributors to the past problem. Even if the beneficiaries today are the same as those in the past, it is not clear that today's markets for taxed products will permit the same percentage of cost pass-throughs that may have been possible in the 1950s and 1960s, when U.S. prices were not so dependent on international forces. If less of the cost is being passed on to consumers, the burden of the Superfund excise taxes will fall in part on current shareholders of the companies involved. These may be an entirely different people than those who benefited from the disposal practices in years gone by.

This apparent inequity was of no particular concern to Congress in 1980. Congress was more concerned then with finding a reliable and politically viable source of income to finance the Superfund cleanups. A bias toward loading costs onto easy targets, with little regard for culpability, is also evident in the Superfund provisions for joint and several liability (which make one party liable for all clean-up costs if other parties are difficult to locate). The Superfund amendments of 1986 do redress some of the potential inequities by adding the broad-based manufacturers tax, although the feedstock tax still contributes roughly 50 percent of SARA's funding.

The liability and distributional issues surrounding Superfund raise questions of both efficiency and equity. The current program creates incentives to increase the cost and to delay the onset of cleanups.[59] At the same time, clean-up actions provide benefits to a broad cross section of the population. But the program is financed primarily by taxes on a narrow subset of all hazardous waste producers. How might these shortcomings be remedied? Both problems suggest that the Superfund program might benefit by a shift in perspective as to who is responsible and who should pay. Real environmental gains might result from a clean-up program financed entirely by general tax revenues, or some other broad-based revenue source, with no assignment of private liability for clean-up costs.[60] Such a system would recognize the equity implications of Superfund cleanups and would reduce the need to expend large sums to force private actions. In addition, it would lower the current barriers that keep firms from coming forward with information concerning past disposal sites and to admit responsibility for them. On the other hand, it would strike many as grossly unfair that firms known to have contributed to problems at certain sites would bear no share of the clean-up costs. A less radical, and budgetarily more feasible, solution might involve an increase in the percentage of the fund that comes from general revenues. More cost recovery from private firms might also be possible if firms were rewarded with reduced penalties for willingly and quickly coming forward with data on their past disposal practices.

Other alternatives for reducing the legal and transaction costs associated with the current Superfund program should be investigated, for these costs stand as a significant barrier to efficient and effective implementation of a national clean-up program. One such alternative may be exemplified by a nonprofit organization formed under the name Clean Sites, Inc. Developed through a working partnership of industry, environmental, and governmental officials, Clean Sites has been designed as an intermediary to negotiate and speed the design and implementation of private clean-up actions. Its activity is not intended to duplicate that of the EPA, but rather to focus on sites that may not in the near future see clean-up work begun because of legal complications. Clean Sites has not been in operation long enough to establish a track record, but it may play a constructive role in addressing the problem wastes. Incidentally, it too has faced problems of liability insurance, and had to seek indemnification from the EPA for responsibility at sites that are remedied through a Clean Sites action.

Where Will Our Wastes Go?

The 1984 RCRA was designed to ensure that by the early 1990s most hazardous wastes generated in the United States will not be disposed of in landfills, surface impoundments, or other forms of land disposal. Where will they go, then? Clearly, the authors of RCRA hoped that waste generators would change production techniques and alter the mix of intermediate inputs to minimize the generation of wastes.[61] This will happen, but wastes will still be generated because it is impossible to eliminate them altogether, and because it becomes increasingly more expensive to reduce their volume as less waste is produced.[62] Other alternatives include the physical or chemical alteration of wastes to render them relatively harmless, injecting them deep into wells, or incinerating them on land or at sea. No doubt some mix of all of these will form the hazardous waste disposal picture of the future, particularly as new technologies emerge and current kinks in the available techniques are worked out.

Incineration of hazardous wastes, particularly at sea, would appear to be an economically viable and relatively safe method of disposal, and might be encouraged at least as an interim method.[63] The current tendency to subject incineration to extensive regulatory scrutiny may be justified in order to avoid future risks, but should not be allowed to result in a lack of disposal capacity.

Other regulatory programs or initiatives also act to constrain or limit the range of disposal alternatives. Moreover, the problem does not solely concern the disposal of currently generated wastes. There is evidence that Superfund cleanups have been delayed as a result of inadequate disposal capacity. Waste minimization efforts will have no effect on existing wastes, although on-site treatment of these wastes may in the long run reduce some pressure on landfilling.

Aside from the relative merits or costs of making the transition from land-based disposal to other methods, we should recognize the existence of a nonregulatory, noneconomic barrier to such transitions—public resistance and fear, which may well override the good intentions of Congress and the EPA and impede any orderly transitions. Further, no one wishes to live or work near hazardous waste disposal sites and activities. Such locally undesirable land uses, dubbed LULUs,[64] are perceived to impose significant costs on those nearby in the form of public health risks or environmental disamenities.

In the case of hazardous waste disposal sites, public fears about health and environmental risks from even the safest facility are so great that the standard approaches to siting may not be enough. The problem is already acute. Very few new disposal facilities have been sited in this country in the last several years, even in the face of increasing demands for safe sites that would operate under permits. Public opposition to hazardous waste facilities is not limited to landfill operations, but extends to incinerators and at-sea burning. The problem is complicated by the control over land-use decisions exerted by state and local governments. The recent trend is toward more rather than fewer restrictions on these waste disposal options.

A growing body of literature is focusing on creative solutions to the siting of hazardous waste management facilities. Consensus-building techniques, economic payoffs or incentives offered to the affected public, alternative forms of compensation (such as parks or other environmental amenities), compensation or insurance pools, as well as other mechanisms are being explored to overcome public reluctance. The success rate to date is not inspiring, and much more needs to be done to avoid a shortfall in disposal capacity, but these techniques give some reason for hope. The need for increased research on and analysis of the risks and costs of hazardous waste disposal is critical, not only to provide a sound basis for regulatory decision making, but also to ensure an informed public. Meeting this need will go a long way toward facilitating many of the other siting mechanisms now being evaluated.

CONCLUSIONS

The hazardous waste problem faced by the United States is a by-product of an advanced economy and technological life-style. It is also a result of our other environmental laws, which have redirected health and environmental risks from the air and water to the land.[65] Coping with the risks of carelessly discarded wastes, while trying to maintain the other aspects of environmental quality and economic well-being, will require a flexible regulatory structure that tries to minimize the costs of transition between the status quo and the future. Our current hazardous waste laws are not designed to do this very well.

In attempting to solve both past and present disposal problems, we incur a real and costly probability of further redistributing risks in unknown directions. Our state of knowledge concerning the benefits and costs of hazardous waste control is shockingly poor. An orderly process of using our experience in regulating other environmental risks, responding to acute and current risks, and leaving other risks alone until better information is available may be the most cost-effective response in the long run. As we have seen in this chapter, however, the tide is running in the other direction. Public attitudes and perceptions concerning hazardous wastes have led to a regulatory response that fails to adequately consider variations in risks and costs and that overlooks key economic incentives for achieving efficient control. More generally, the current emphasis on industrial hazardous wastes may have the larger side-effect of diverting attention and resources from environmental issues involving greater risks.

This last point is easily illustrated. The tendency to focus on hazardous waste as an industrial problem ignores the potential contribution of other sources to the hazardous waste problem. Yet, as we are finding with other forms of pollution, hazardous waste risks may exist closer to home, in fact within the home. Paint thinners, used motor oil or gasoline, and many other common household materials become hazardous wastes when they are no longer needed. Improper storage, handling, and disposal of these products may in some cases pose greater health risks to the public than industrial wastes. Yet our current laws do not subject such wastes to careful regulation. Nor do we know very much about the quantities of household hazardous wastes that exist, and how they are handled. The potential risk may be significant. In many ways the problem is like that of non-point source water pollution discussed in chapter 4 or indoor air pollution discussed in chapter 3. These frontier environmental issues present many of the same difficult and uncertain choices that are associated with hazardous waste management. Will we learn from the past? The lessons from our current hazardous waste regulations do not give much reason for confidence. Yet these regulatory programs are in their relative infancy, and there is always the hope they can be made more efficient and effective.

NOTES

1. Based on responses to polls conducted by the Roper Organization in December 1987 and January 1988.

2. This figure is based on the EPA's hazardous waste and Superfund outlay estimates for 1989, as reported in Environmental Protection Agency, *Summary of the 1990 Budget* (January 1989).

3. Resource Conservation and Recovery Act, 42 U.S.C. section 1004 (5) 6903; ELR STAT. 42006.

4. 40 C.F.R., section 261. For further discussion of source of these refinements, see below.

5. The question of when a material is a solid waste is almost as complex as defining whether a solid waste is hazardous. For an interesting discussion of solid waste definitions, see B. Farelick, "When Is a Waste Not a Waste," *Environmental Forum* vol. 4, no. 5 (September 1985) pp. 26–32.

6. J. E. McCarthy and M. E. Anthony Reisch, *Hazardous Waste Fact Book*, 87-56 ENR (Washington, D.C., Congressional Research Service, 1987).

7. Congressional Budget Office, *Hazardous Waste Management: Recent Changes and Policy Alternatives* (Washington, D.C., Government Printing Office, 1985) p. 26.

8. Although the survey data underlying this remarkable statistic are not available, they do apply to all wastes treated on or off site. The 10 facilities, however, appear to be associated directly with manufacturing activities and are not commercial TSDFs. For a discussion of the waste management industry and alternative data sources, see McCarthy and Reisch, *Hazardous Waste Fact Book*.

9. General Accounting Office, *Superfund: Extent of Nation's Potential Hazardous Waste Problem Still Unknown*, GAO/RCED-88-44 (December 1987).

10. Office of Technology Assessment, *Superfund Strategy* (Washington, D.C., Government Printing Office, 1985).

11. Congressional Budget Office, *Hazardous Waste Liabilities at Federal Facilities* (Washington, D.C., forthcoming).

12. Universities Associated for Research and Education in Pathology, *Health Aspects of the Disposal of Waste Chemicals* (Bethesda, Md., 1985).

13. Environmental Protection Agency, "Preliminary Benefit-Cost Analysis for Regulating VOC Emissions from Treatment, Storage and Disposal Facilities Managing RCRA Waste," Economic Analysis Branch, Office of Air Quality Planning and Standards, EPA, July 1985.

14. Temple, Baker, and Sloan, Inc., "Regulatory Impact Analysis: Proposed Standards for the Management of Used Oil," prepared for the Office of Solid Waste, Environmental Protection Agency, July 1985.

15. Environmental Protection Agency, "Unfinished Business: A Comparative Assessment of Environmental Problems," overview report, Office of Policy Analysis, February 1987.

16. 40 C.F.R., section 261.30, 1988.

17. 40 C.F.R., section 261.24, 1988.

18. Office of Technology Assessment, *Technologies and Management Strategies for Hazardous Waste Control* (Washington, D.C., Government Printing Office, 1983).

19. J. Quarles, *Federal Regulation of Hazardous Wastes: A Guide to RCRA* (Washington, D.C., Environmental Law Institute, 1982).

20. General Accounting Office, "Illegal Disposal of Hazardous Wastes: Difficult to Detect or Deter," GAO/RCED-85-2, 1985.

21. 42 U.S.C., sections 3004, 6924; ELR STAT. 41909.

22. The congressional Office of Technology Assessment has identified criteria for establishing equivalence: the state regulations must (1) control substantially the same universe of wastes in each state; (2) provide adequate regulatory authority to control generators, transporters, and operators of TSDFs; and (3) demonstrate adequate funding and personnel for administration and enforcement of the program. See Office of Technology Assessment, *Superfund Strategy*.

23. 42 U.S.C., sections 6901–6987; ELR STAT. 41901.

24. One congressman who played a key role in designing the 1984 RCRA amendments has provided an interesting account of the rationale behind the new law. See J. J. Florio, "Congress as Reluctant Regulator: Hazardous Waste Policy in the 1980s," *Yale Journal on Regulation* vol. 3, no. 2 (1986) pp. 251–382.

25. 42 U.S.C., sections 3001, 6921; ELR STAT. 42012.

26. Office of Technology Assessment, *Technologies and Management Strategies.*

27. 42 U.S.C., section 3004.

28. Freeman, A., "Firms Curb Hazardous Waste to Avoid Expensive Disposal," *Wall Street Journal,* May 31, 1985.

29. Environmental and Energy Study Institute, *The National Hazardous Waste Land Disposal Restrictions: Better Data, Clearer Policy Needed to Make It Work* (Washington, D.C., 1987).

30. Environmental Law Institute, *Six Case Studies of Compensation for Toxic Substances Pollution: Alabama, California, Michigan, Missouri, New Jersey, and Texas,* report prepared for the Congressional Research Service (Washington, D.C., Government Printing Office, 1980).

31. 42 U.S.C., sections 9601–9657; ELR STAT. 41941.

32. A useful guide to SARA and the history of its development can be found in Environmental Law Institute, *Superfund Desk Book* (Washington, D.C., Environmental Law Institute, 1986).

33. For an empirical analysis of the EPA's remedial action decision process, see H. S. Barnett, "The Allocation of Superfund, 1981–1983," *Land Economics* vol. 61, no. 3, (1985) pp. 255–262.

34. Environmental Protection Agency, *Superfund Advisory* (Washington, D.C., Winter 1989).

35. See Office of Technology Assessment, *Superfund Strategy.*

36. A detailed appraisal of the Superfund natural resource damage process and the role of economics in these procedures can be found in Raymond J. Kopp and V. Kerry Smith, eds., "Valuing Natural Assets: The Economics of Natural Resource Damage Assessments," Resources for the Future, manuscript.

37. Included are settlements on sites at which private parties conduct all the work, as well as on sites where the EPA is reimbursed by private parties. See Environmental Protection Agency, *Superfund Advisory.*

38. Environmental Protection Agency, *Superfund Advisory.*

39. See Environmental Law Institute, *Superfund Desk Book,* p. 4.

40. R. L. Raucher, "The Benefits and Costs of Policies Related to Groundwater Contamination," *Land Economics* vol. 62, no. 1 (1986) pp. 33–45.

41. Congressional Budget Office, *Hazardous Waste Management.*

42. McCarthy and Reisch, *Hazardous Waste Fact Book.*

43. Congressional Budget Office, *Hazardous Waste Management,* p. 53, note 6.

44. See Environmental Protection Agency, *Summary of the 1990 Budget.*

45. Environmental Protection Agency, *Case Studies—Remedial Response at Hazardous Waste Sites,* EPA-540-540/2-84-002, Office of Emergency and Remedial Response, EPA (Washington, D.C., 1984).

46. See General Accounting Office, *Superfund: Extent of Nation's Potential Hazardous Waste Problem,* p. 23.

47. Congressional Budget Office, *Hazardous Waste Liabilities.*

48. Environmental Protection Agency, "Hazardous Waste Assessment—A Progress Report," briefing materials, 1984.

49. J. Evans, C. Petito, and D. Gravellese, "Cleaning Up the Gilson Road Hazardous

Waste Site," Discussion Paper no. E-86-03, Kennedy School of Government, Harvard University, February 1986.

50. Paul R. Portney, "Reforming Environmental Regulation: Three Modest Proposals," *Columbia Journal of Environmental Law* vol. 13 (1988) pp. 501–515.

51. David Harrison, Jr., and coauthors, "Research and Demonstration of Improved Methods for Carrying Out Benefit-Cost Analysis of Individual Regulations," vol. 1, submitted to the Environmental Protection Agency under Cooperative Agreement no. 68-809-702-01-0, November 1984; V. Kerry Smith and coauthors, "Valuing Changes in Hazardous Waste Risks: A Contingent Valuation Analysis," vol. 3, submitted to the Environmental Protection Agency under Cooperative Agreement no. CR 811075, February 1985.

52. A useful overview of recent work on understanding of public perceptions toward risk is found in P. Slovic, B. Fischhoff, and S. Lichtenstein, "Why Study Risk Perception?" *Risk Analysis* vol. 2, no. 2 (1983) pp. 83–93.

53. See especially *Injuries and Damages from Hazardous Wastes—Analysis and Improvement of Legal Remedies,* S. Rept. 97-12, 97 Cong. 2 sess. (1982) (Washington, D.C., Government Printing Office, 1982); J. Trauberman, *Statutory Reform of "Toxic Torts": Relieving Legal, Scientific, and Economic Burdens on the Chemical Victim* (Washington, D.C., Environmental Law Institute, 1983).

54. A wide range of topics associated with insurance and Superfund are discussed in *Insurance Issues and Superfund,* Committee Print 99-61, Senate Committee on Environment and Public Works, 99 Cong. 1 sess. (1985); Fred Smith, "Beyond Superfund," *Wall Street Journal,* October 5, 1984.

55. A congressional analysis has found that most of the 100,000 companies that deal with hazardous wastes are unable to find insurance. General Accounting Office, "Hazardous Wastes Issues Surrounding Insurance Availability," GAO/RCED-88-2, 1988.

56. For an insightful analysis of Superfund litigation problems and their costs, see A. Light, "A Defensive Counsel's Perspective on Superfund," *Environmental Law Reporter* vol. 15, no. 7 (July, 1985); 15 ELR 10203.

57. Cost recovery data has been derived from Environmental Protection Agency, *Superfund Advisory,* and budget data are contained in Congressional Budget Office, *Environmental Federalism: Allocating Responsibilities for Environmental Protection* (Washington, D.C., September 1988).

58. The EPA administrator proposed to do just that in a notice of rulemaking signed on August 16, 1985.

59. Between 1980 and 1987, 444 private cleanups were undertaken, at a total cost of more than $642 million. See Environmental Protection Agency, *Environmental Progress and Challenges: EPA's Update* (Washington, D.C., August 1988).

60. This proposal does not address the issue of liability for personal injury associated with a hazardous waste site. Although there have been few such cases to date, the fear of compensation claims by victims appears to be a significant impediment to the firms' acceptance of liability for clean-up costs.

61. National Academy of Sciences, *Reducing Hazardous Waste Generation: An Evaluation and a Call for Action* (Washington, D.C., National Academy Press, 1985).

62. There is growing interest in the reduction or recycling of hazardous wastes as an alternative to disposal. Part of the interest stems from a perceived bias in RCRA against nondisposal technologies (see, for example, R. G. Gordon, "Legal Incentives for Reduction,

Reuse, and Recycling: A New Approach to Hazardous Waste Management," *Yale Law Journal* vol. 95, no. 4 (1985) pp. 810–831). The increasing costs of waste disposal would seem to be a sufficient incentive for firms to explore waste minimization options with little encouragement. However, the evidence either way is limited. One recent survey found that wastes have been reduced by more than 20 percent in the chemical industry over the period 1981–1985.

63. Environmental Protection Agency, *Assessment of Incineration as a Treatment Method for Liquid Organic Hazardous Wastes* (Washington, D.C., 1985).

64. F. J. Popper, "Siting LULU's," *Planning—APA* vol. 47, no. 7 (April 1981) pp. 12–15.

65. For an interesting account of how government programs may act to redistribute environmental and health risks, see D. Whipple, "Redistributing Risks," *Regulation* (May/June 1985) pp. 37–44.

six

Toxic Substances Policy

Michael Shapiro*

INTRODUCTION

The decades following World War II saw a dramatic expansion in the role of chemicals, especially man-made (or synthetic) chemicals, in the economy of the United States. Chemical technology made possible significant advances in industrial materials, fabrics, agricultural productivity, electronics, and a diverse array of other fields. But as time progressed, the public and policymakers became increasingly concerned that these benefits were being obtained without due consideration for the possible adverse consequences associated with chemical production and use. Incidents involving specific environmental and human health problems associated with particular chemicals, a growing concern that chronic diseases such as cancer might be associated with environmental (although not necessarily chemical) causes, and a concern over the comparative lack of data on many chemicals led to a demand for federal regulatory activities in the areas of chemical manufacturing and use. This chapter describes the two major legislative acts passed as a result of these concerns—the Toxic Substances Control Act (TSCA) and the Federal Insecticide, Fungicide, and Rodenticide Act (FIFRA)—and examines some of the significant policy issues that have arisen in their implementation.

*The interpretations and opinions presented in this chapter are those of the author and do not necessarily reflect the views of the U.S. Environmental Protection Agency.

Over the past fifteen years Congress has enacted and the Environmental Protection Agency (EPA) has begun to implement a framework for assessing and regulating the risks associated with major classes of commercial chemical substances. In general, these activities focus on the control of toxic substances, which might be loosely defined as substances capable of causing adverse human health or environmental effects under anticipated conditions of exposure (see also chapter 5). However, the sum total of all federal regulation of toxic substances encompasses a much broader scope of activity than is considered in the present chapter. All of the other environmental laws discussed in this book have major components that deal with what we would consider to be toxic substances, as do the Occupational Safety and Health Act, the Consumer Product Safety Act, and other statutes summarized in table 6–1.

The concept of toxic substances, as described above, encompasses a broad array of possible human health and environmental effects.[1] Some acute health effects occur fairly rapidly upon exposure to certain levels of toxic materials—skin or lung irritation, for example. Currently, exposures likely to give rise to acute effects will occur only under abnormal circumstances, such as leaks or spills. Thus, at least until the recent Bhopal tragedy, the EPA's attention has been focused primarily on effects that may result from more routine exposures that are below levels which would manifest acute effects. Such lower-level exposures are often characterized by significant latency periods between the initiation of exposure and the manifestation of the toxic effects; the toxic effects may include carcinogenicity, reproductive problems, neurotoxicity, and chronic organ toxicity (for example, liver disease).

The latencies associated with several of these effects—combined in some cases with the relatively low levels of exposure, exposure to multiple sources of potential harm, and incomplete historical records of either chemical-specific exposures or disease incidence—make it difficult to attribute various diseases either to specific substances or to toxic substances in general.

Only cancer has been studied extensively. For this disease, comparisons across different geographic and cultural groups suggest that a large majority of cancer cases are "avoidable" in the sense that they are due to various life-style or environmental factors that could be substantially changed or modified. Doll and Peto have estimated that 75 to 80 percent of all cancers are potentially avoidable in this sense.[2] Their conclusion does *not* mean, however, that the majority of cancers are caused by exposure to chemical substances that might be regulated by the EPA. Indeed, these researchers estimated that only 2 to 8 percent of avoidable cancer deaths were attributable to occupation, less than 1 to 5 percent to pollution, and less than 1 to 2 percent to industrial products. They also concluded:

> Examination of the trends in American mortality from cancer over the last decade provides no reason to suppose that any major new hazards were introduced in the preceding decades, other than the well-recognized hazard of cigarette smoking, which has extended from men to women, and the cause (whatever it may be) of the increase in melanoma (a skin cancer).[3]

These conclusions have been controversial; the authors themselves admit to considerable uncertainty in their estimates and point out that because of latency periods in developing cancer, and the rapid rise of chemical production from 1950 to 1970, it may be too soon to detect patterns of occupational or product-related cancers stemming from this industrial activity. Moreover, even if the basic conclusions of these authors are correct, the absolute numbers of cancer deaths implied by their percentages would justify serious regulatory attention. (For instance, 3 percent of the 416,500 cancer deaths in the United States in 1980 is still 12,500 deaths.) Moreover, the causes of other chronic adverse health effects are not even as well understood as those of cancer,[4] justifying concern that chemical products may be associated with some portion of these hazards.

Under the balancing approach discussed in chapter 2, a rational model for managing toxic substances would involve the following steps:

1. Identification of adverse effect(s)
2. Estimation of the relationship between exposure to hazardous substances and the response for effects of concern
3. Evaluation of current exposures to the chemical, including level of exposure, duration, frequency, routes of exposure (oral, dermal, for example) and other relevant factors
4. Combining of exposure and dose-response data to predict the *risks* associated with the chemical
5. Identification of possible options for reducing risks
6. Evaluation of the costs and impacts of each risk reduction option and the degree of risk reduction achieved relative to the situation that would exist if no action is taken
7. Selection of an appropriate option, which may include a decision not to control, based upon a comparison of risk reduction and costs

Unlike some other environmental laws, which require decisions to be based upon technological availability or margins of safety, and which limit consideration of economic factors, both TSCA and FIFRA encourage this risk-cost balancing approach to decision making and take a comprehensive

Table 6–1. Federal Laws Dealing with Toxic Substances

Statute	Responsible agency	Sources covered
Toxic Substances Control Act	EPA	Requires premanufacture evaluation of all new chemicals (other than food, food additives, drugs, pesticides, alcohol, tobacco); allows EPA to regulate existing chemical hazards
Clean Air Act	EPA	Hazardous air pollutants
Clean Water Act	EPA	Toxic water pollutants
Safe Drinking Water Act	EPA	Drinking water contaminants
Federal Insecticide, Fungicide, and Rodenticide Act	EPA	Pesticides
Section 346(a) of the Food, Drug, and Cosmetic Act	EPA	Tolerances for pesticide residues in human food and animal feeds
Resource Conservation and Recovery Act	EPA	Hazardous wastes
Maritime Protection, Research, and Sanctuaries Act	EPA	Ocean dumping
Comprehensive Environmental Response, Compensation, and Liability Act of 1980	EPA	Hazardous waste releases
Food, Drug, and Cosmetic Act	FDA	Basic coverage of food, drugs, cosmetics
Food additives amendment	FDA	Food additives
Color additives amendments	FDA	Color additives
New drug amendments	FDA	Drugs
New animal drug amendments	FDA	Animal drugs and feed additives
Medical device amendments	FDA	Medical devices
Fair Packaging and Labeling Act	FDA	Packaging and labeling of food and drugs for humans or animals, and of cosmetics and medical devices

Occupational Safety and Health Act	OSHA	Workplace toxic chemicals
Federal Hazardous Substances Act	CPSC	"Toxic" household products (equivalent to consumer products)
Consumer Product Safety Act	CPSC	Dangerous consumer products
Poison Prevention Packaging Act	CPSC	Packaging of dangerous children's products
Lead-Based Paint Poison Prevention Act	CPSC	Use of lead paint in federally assisted housing
Hazardous Materials Transportation Act	DOT (Materials Transportation Bureau)	Transportation of toxic substances generally
Federal Railroad Safety Act	DOT (Federal Railroad Admin.)	Railroad safety
Ports and Waterways Safety Act	DOT (Coast Guard)	Shipment of toxic materials by water
Dangerous Cargo Act		Shipment of toxic materials by water
Federal Meat Inspection Act	USDA	Food, feed, and color additives and pesticide residues in meat and poultry products
Poultry Products Inspection Act	USDA	
Egg Products Inspection Act	USDA	Egg products
Federal Mine Safety and Health Act	MSHA	Coal mines or other mines

EPA = Environmental Protection Agency; FDA = Food and Drug Administration; OSHA = Occupational Safety and Health Administration; CPSC = Consumer Product Safety Commission; DOT = Department of Transportation; USDA = Department of Agriculture; MSHA = Mine Safety and Health Administration

Sources: Toxic Substances Strategy Committee, *Toxic Chemicals and Public Protection* (Washington, D.C., Government Printing Office, 1980); Council of Environmental Quality, *Environmental Quality—1982* (Washington, D.C., Government Printing Office, 1982).

view of risk management across environmental media. Moreover, significant parts of both pieces of legislation provide mechanisms for the EPA to gather the data necessary to make such decisions. This may be the feature that most differentiates TSCA and FIFRA from the other environmental statutes discussed in this book.

Another important distinction is that both TSCA and FIFRA deal with the way in which chemicals and pesticide products are produced and/or used, rather than focusing solely on residuals generated in the course of manufacturing them. For example, consumer exposure to a toxic solvent used in a spray paint is inherent in the use of the paint in its intended application. In such circumstances the only feasible way to significantly reduce the risk from exposure may be to eliminate or restrict the use of that paint in certain applications. Thus, both TSCA and FIFRA enable the Environmental Protection Agency to ban the use of specific chemicals. On the other hand, once the risks of a chemical use are known, users may react to this knowledge by choosing other products; manufacturers may react to such market changes, as well as to liability considerations, by reformulating products to replace the toxic material.

Such market reactions can be extremely powerful. Manufacturers virtually eliminated the use of several toxic solvents from consumer products once information on their potential toxicity became widely known and the EPA initiated a regulatory investigation. In other cases the response may not be as rapid. The point is that the nature of product regulation expands the potential risk-management options from the normal array of technological controls, like those discussed in the preceding chapters, to the opposite extremes of bans and hazard-communication strategies.

In the next section of this chapter, the background discussion includes a brief history of the chemical industry, particularly its development in the United States and its current economic importance. We then examine some of the health and environmental problems that became concerns as the post–World War II growth of the industry exposed workers and the general population to a vast array of biologically active materials, both synthetic and natural. These concerns led to a regulatory response during the 1970s, when TSCA was passed and federal pesticide policy was restructured. In the third section, TSCA and FIFRA are described, their similarities are identified, and their significant differences contrasted.

The fourth section discusses the progress made, problems encountered, and significant policy issues which have arisen in the implementation of chemical product laws. The discussion focuses primarily on TSCA, and is organized around three major themes implicit or explicit in the law: the provision of adequate data on chemicals, the control of unreasonable risk from existing chemicals, and the need to balance risk reductions against the benefits from continued innovation of new chemical substances. The final section offers conclusions and policy recommendations.

BACKGROUND

The Chemical Industry in the Economy

Historians of the modern chemical industry have tied its origins to the beginnings of the industrial revolution in the late 1700s.[5] Through most of the nineteenth century the industry's fortunes were based on the production of relatively simple inorganic chemicals for the manufacture of basic materials such as soap, glass, cotton textiles, and paper, as well as for agricultural use as fertilizers.[6] A major revolution in the industry occurred in 1856 when an Englishman, W. H. Perkins, developed the first synthetic dye, mauve. This discovery was rapidly followed by a host of similar synthetic dyes which greatly enhanced the colors available to textile makers. The era of synthetic organic chemicals had begun. Subsequent developments, including pharmaceuticals, photographic chemicals, film, and rubber additives all had their roots in the technology of dyes.

The era between the two world wars saw the commercialization of a second revolutionary technology, synthetic polymers. Polymers are large molecules made up of repeating chains of basic, building-block molecules linked through chemical bonds. Polymers such as vulcanized rubber and celluloid, derived from natural materials, and rayon, derived from natural sources of cellulose, were available during the nineteenth century. And in the early 1900s Bakelite, the first synthetic polymer to gain industrial importance, was introduced. However, it was not until after World War I that fundamental breakthroughs in the understanding of polymeric structures and polymerization led to a myriad of new basic substances. These new materials made possible applications in plastics, coatings, adhesives, and synthetic rubber products. In fact, the majority of polymer families of commercial importance even today were first developed during this period by I. G. Farben in Germany, Imperial Chemical Industries (ICI) in England, and du Pont in the United States. The commercial potential of synthetic polymers was further enhanced by the introduction during this period of inexpensive petroleum-based feedstocks.

After World War II, the U.S. chemical industry grew rapidly, benefiting from the overall economic expansion of the economy, substitution of synthetic for natural materials, and the demand for new products made available by the chemical technologies. The growth was led by the rapid

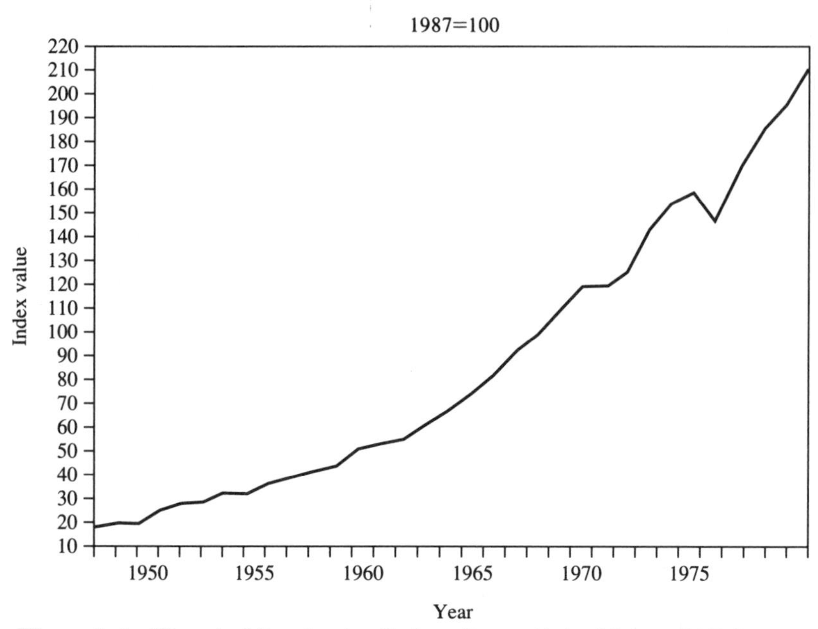

Figure 6–1. Chemical Production Index. *Source: Federal Reserve Bulletin.*

expansion of polymeric materials into fabrics, manufactured products, coatings, adhesives, and various industrial applications. This growth can be illustrated by data on the value added by different manufacturing industries in the United States. In 1899 the chemical industry accounted for 4.3 percent of all value added in manufacturing. By 1970 this share had more than doubled, to 9.3 percent. Figure 6–1 illustrates the relative growth of the U.S. chemical industry since World War II, using an index of total production. While all this may sound abstract, such essential products as nylon fueled this growth and provided some truly remarkable breakthroughs in the quality of people's lives.

In 1986 the U.S. chemical industry accounted for about $214 billion of sales (see table 6–2), making it the world's largest chemical industry.[7] Other major chemical-producing countries include West Germany, the United Kingdom, France, Italy, and Japan, all of whom had 1986 sales exceeding $25 billion. The international industry is extremely competitive. Despite the large size of the U.S. chemical industry, American firms enjoy no particular size advantage over their foreign competitors. Indeed, measured in terms of chemical sales, a majority of the top ten chemical firms are foreign. United States chemical firms have enjoyed a historical

Table 6–2. Profile of the U.S. Chemical Industry, 1986 (except as noted)

Sales (billions of $)	214.0
Exports (billions of $)	22.8
Imports (billions of $)	15.0
No. of companies (1982)	7,527
No. of establishments (1982)[a]	11,901
No. of chemical substances[b]	60,000+
Employment (thousands)	1,027
R&D expenditures ($ billions)	9.1
Capital expenditures ($ billions)	16.75

Note: Statistics are based on the Standard Industrial Classification (SIC) 2800, chemicals and allied products.

Sources: "Facts and Figures for the Chemical Industry," *Chemical & Engineering News* vol. 65 (June 8, 1987) pp. 24–76; Department of Commerce, *1982 Enterprises Statistics: Annual Report on Industrial Organization*, ES82-1 (Washington, D.C., Government Printing Office, 1984); Department of Commerce, *1982 Census of Manufacturers: General Summary*, MC82-S-1 (Washington, D.C., Government Printing Office, 1983).

[a] An establishment is a business unit at a single physical location (that is, a plant); a company is a business organization consisting of all establishments under common ownership or control. Note that some non-chemical companies own one or more chemical plants; therefore the number of establishments does not reflect solely the plants owned by the 7,527 chemical companies.

[b] This is an approximate figure based on the number of chemical substances on the TSCA inventory. The inventory excludes the active ingredients in pesticides, as well as food additives and drugs, but the overall count is probably close to this figure.

trade advantage, however, resulting from both the availability of cheap petrochemical feedstocks and their technological advantages in certain market areas. As a result, U.S. trade in chemicals has consistently yielded a net surplus of exports over imports. This surplus exceeded $7 billion in 1985. But several factors suggest that it will be difficult to maintain this surplus in the future, and that the decline might continue. These factors include the increasing relative cost of domestic feedstocks and the establishment of petrochemical manufacturing capacity in countries such as Canada, Mexico, and Saudi Arabia, where low-cost feedstocks are available.

In response to changing world trade conditions, U.S. firms are turning increasingly to new-product development, utilizing advanced technologies to maintain their competitive positions. Thus there is emphasis throughout the industry on the development and commercialization of new technologies in areas such as biotechnology, catalysis, ceramics, membranes, and specialized high-performance polymers that have applications in rapidly growing industries like electronics and pharmaceuticals or where significant potential exists to substitute for other materials, as in automobile and aircraft manufacturing.[8]

Pesticides

The history of the pesticide industry follows a similar pattern. Broadly defined, pesticides include "chemicals used in agriculture to protect, preserve and improve crop yields, excluding fundamental nutrients,"[9] as well as chemicals used to control disease vectors, protect materials from biological attack, maintain rights-of-way, protect gardens, and enhance aesthetic conditions around homes and other settings. Pesticides were used to some degree in the nineteenth century, but the materials used were primarily based on sulfur and arsenic, plant extracts, and crude petroleum fractions.[10] However, as with polymers, a major technological revolution occurred between the two world wars, with the introduction of the first synthetic organic insecticides. Insecticides such as DDT, based on chlorinated organic compounds, proved to be remarkably effective against a broad range of pests and provided relatively long-term effectiveness. As Rachel Carson pointed out, however, the lack of specificity and the persistence of these pesticides were later found to be significant environmental disadvantages.

The new organic pesticides proved especially effective during World War II in controlling the spread of insect-borne disease. After the war, production of organic pesticides accelerated as their application in agriculture became widespread. In addition to the original chemicals based on chlorinated organics, research yielded other families and derivatives with better specificity to particular pests, higher potency to target organisms, and reduced persistence in the environment. (Some of these also resulted in health and environment concerns, however.) New applications were also developed. The current most rapidly growing application of pesticides in the United States involves herbicides—chemicals used to control weeds. The introduction of new herbicides has made possible an increase in crop yields while reducing the need for tillage, resulting in both labor savings and reductions in soil erosion. As of 1986, the total annual value of pesticide shipments was more than $5 billion, with net exports of about $1.3 billion.

Research continues to be directed toward developing pesticides that are more effective and specific. Some pesticides are based on mechanisms other than direct toxicity. For example, chemicals designed to duplicate sexual attractants of specific insect species can be used to confuse insects, making mating less likely. Biotechnology—including the possible bioengineering of plants to improve pest resistance and tolerance to herbicides—

is expected to contribute significantly in the future to new pesticide technologies.

Health and Environmental Concerns

The explosive growth of the chemical industry after World War II clearly provided substantial economic benefits. At the same time, that growth resulted in the rapid increase in production and dispersion into the environment of natural and synthetic toxic materials. As it turned out, our use of these materials in some cases preceded our ability to adequately forecast their consequences for health and the environment. During the nineteen-fifties, sixties, and early seventies, a series of events led many to believe that existing policies were not capable of ensuring the safety of chemical products:

1. Synthetic pesticides, such as DDT, became distributed widely in the environment, and, due to their ability to persist and accumulate in fatty tissue, were found in elevated levels in higher organisms, including the tissues of humans. Evidence mounted on the adverse effects of these materials in the environment and on human health.

2. Between 8 and 11 million workers were estimated to have been exposed to asbestos in the United States since the beginning of World War II. Associations between asbestos exposure and cancer were reported in the 1950s, and subsequent studies have further documented this risk.[11]

3. Vinyl chloride, an important chemical intermediate, was shown to be responsible for a rare form of liver cancer.[12]

4. In Japan, widespread neurotoxic effects occurred when organic mercury compounds, resulting from an unanticipated environmental transformation of mercury wastes, contaminated seafood.[13]

5. About 1.2 billion pounds of polychlorinated biphenyls (PCBs) were manufactured and sold between 1930 and 1975. These chemicals were associated with a variety of health effects, including carcinogenicity in laboratory animals.[14]

6. Production and improper handling of the pesticide Kepone at Hopewell, Virginia, led to a number of serious health effects among employees, and resulted in the closing of 100 miles of the James River and its tributaries to commercial fishing.[15]

7. The flame retardant Tris, used to meet flammability standards on children's sleepwear, had to be banned following the development of information identifying the chemical as a potential human carcinogen.[16]

8. In Michigan, toxic polybrominated biphenyl (PBB) flame retardants were inadvertently mixed with animal feed. Cows and chickens throughout

the state consumed contaminated feed and ultimately had to be destroyed, but not before contaminated products had already reached the public.[17]

In addition to these and other specific events, there was growing general concern over the rise of cancer death rates and the linkage that some claimed between these rates and the growth in chemical production and use. The empirical support for this linkage, or even for the existence of a rapid increase in age-adjusted cancer rates (other than lung cancer), is subject to differing interpretations. Nevertheless, the growing public concern about cancer contributed significantly to the impetus for toxic substances legislation.

LEGISLATIVE RESPONSE

The Toxic Substances Control Act

Specific toxic chemical disasters, as well as the general concern for environmental protection in the early 1970s, led to increasing interest in legislative action to control the risks associated with commercial chemicals. At this juncture, it would appear that a variety of different approaches could have been taken to address these concerns. At one extreme there might have been a program of registration and approval for all commercial chemicals, analogous to the process for controlling drugs. This would have involved mandatory, comprehensive testing of all chemicals and regulatory approval for their manufacture and distribution. Another approach would have been to amend or enact other statutes associated with air, water, solid waste, consumer safety, and worker protection to cover those aspects of risk associated with chemical toxicity. Many of these laws were, in fact, being modified or enacted around this time.

Congress could also have required labeling, possibly combined with testing requirements, to assure that users were informed about the hazards associated with chemicals and that manufacturers considered the potential liabilities associated with the commercial introduction of their products. Such an approach would rely on the ability of users to make informed judgments about the acceptability of risks posed by the chemicals and of the liability system to internalize the costs of introducing hazardous chemicals. On the other hand, Congress might have done nothing, on the assumption that heightened awareness about toxic chemical hazards would lead to a market solution based on the workings of the product liability system. That is, companies would undertake the necessary testing, labeling, and product control activities in order to limit their liabilities for adverse effects associated with their products.

Ultimately, Congress elected to enact separate toxic chemical regulatory legislation, although, except for pesticides, it stopped considerably short of the drug model. Many factors contributed to this decision, but three appear to have been key: (1) the desire to provide a mechanism for a comprehensive or life-cycle framework for management of chemical risks, simultaneously taking into account risks attributable to product manufacturing and use, and wastes released to the air, water, and land; (2) the desire that such a comprehensive framework be applied to chemicals *before* they went into widespread commercial use; and (3) the perceived need to provide special regulations to ensure that toxicity data would be developed for commercial chemicals. Each of these objectives is addressed in the Toxic Substances Control Act that emerged in 1976 after five years of development and debate.[18]

In passing this act, Congress attempted to provide the Environmental Protection Agency with comprehensive authority to identify, evaluate, and (where appropriate) regulate risks associated with the life-cycle of commercial chemical substances.[19] This authority encompasses chemicals that are already in commerce as well as new chemicals ready for initial commercial manufacture.

The major objectives of the Toxic Substances and Control Act are stated clearly:

> (1) . . . adequate data should be developed with respect to the effect of chemical substances and mixtures on health and the environment[,] and the development of such data should be the responsibility of those who process such chemical substances and mixtures;
>
> (2) . . . adequate authority should exist to regulate chemical substances and mixtures which present an unreasonable risk of injury to health or the environment . . .
>
> (3) . . . authority over chemical substances and mixtures should be exercised in such a manner as not to impede unduly or create unnecessary economic barriers to technological innovation while fulfilling the primary purpose of this act to assure that such innovation and commerce . . . do not present an unreasonable risk of injury to health or the environment.[20]

In providing legislative authorities to implement these objectives, Congress included in TSCA provisions enabling the EPA to gather basic data on chemicals, to require testing by industry, and to regulate risks. In doing so, Congress also distinguished in important ways between new chemicals and existing chemicals. As we will see, however, the new authorities have resulted in less activity than one would have expected when the law was passed.

Existing chemicals. One of the first things the EPA had to do in implementing TSCA was to establish an inventory—referred to here as the TSCA inventory—consisting of chemicals that were either in commercial production or were imported between 1975 and 1977, and chemicals that subsequently began to be imported or produced commercially. More than 60,000 chemicals are on the inventory at present, and these existing chemicals are therefore potentially subject to certain testing or control provisions of TSCA. The EPA can require manufacturers to perform necessary tests to characterize the human and environmental effects associated with specific chemicals or categories of chemicals. The agency can on its own select candidates from the TSCA inventory for testing. However, TSCA provides another mechanism for initiating testing; the law establishes an independent government body known as the Interagency Testing Committee (ITC), which regularly nominates chemicals for testing. When this happens, the EPA must within one year initiate action to require testing or determine that testing is not needed for that chemical.

In order to require that one of the existing chemicals be tested for toxicity, the EPA must determine that existing data are insufficient to determine its effects, that testing is needed to develop such data, and that the chemical may present an unreasonable risk of injury to health or the environment, or that there is substantial production and potential exposure. *It is important to note that the agency must support these findings by going through a full rulemaking process.* In other words, to have a chemical tested the EPA must go through a process similar to the one it uses before setting an ambient air or water quality standard or a new-source performance standard; the process includes public notification of the proposed testing rule, a period for public comment, then issuance of a final testing requirement. This can be an extremely time-consuming and resource-intensive process, taking up to two years even when things go relatively smoothly.

When adequate data about an existing chemical are available, TSCA gives the Environmental Protection Agency authority to regulate its manufacture, processing, use, distribution, or disposal. The regulatory options open to the EPA are very broad, ranging from outright bans on manufacture or specific uses to less draconian labeling requirements. The EPA must support whatever decision it makes by finding that there is a reasonable basis to conclude that the chemical presents or will present an unreasonable risk. Just as it does when requiring that a chemical be tested, the EPA must go through an official rulemaking procedure when imposing such restrictions. For example, the EPA used this authority to ban the use of chlorofluorocarbon propellents in most aerosol products, because of the potential risk of stratospheric ozone depletion.

The EPA's decision to regulate existing chemicals is largely discretionary, but it must initiate regulatory action within 180 days of receiving information indicating that a chemical presents " . . . a significant risk of serious or widespread harm to human beings from cancer, gene mutation or birth defects."

New chemicals. One of the most controversial aspects of TSCA—and a major reason it took so long to be enacted—is the authority it grants the EPA to review new chemicals. Prior to TSCA, some had advocated the testing of all new chemicals for health and environmental effects before commercialization. Others claimed no such review was necessary, except possibly for certain classes of chemicals determined to be potentially hazardous.[21] The compromise reflected in TSCA requires that the EPA be notified at least 90 days before the manufacture of a new chemical substance. Testing is not automatically required, although any test data which may be available on the new chemical must be provided in what is known as the Premanufacture Notice (PMN). The EPA then has a 90-day period (extendable to 180 days) in which to review the proposed startup of production. If data are insufficient to review the chemical adequately, and if the chemical may present an unreasonable risk to health or the environment or will be produced in substantial volume, the agency can ban or limit manufacture until the necessary information is provided. In contrast to testing requirements for existing chemicals, these actions can be taken quickly and impose a relatively low administrative burden. Moreover, the EPA can prevent manufacture of a new chemical until the required data are provided, whereas existing chemicals can continue to be produced and sold during the EPA's development of the testing requirement and the testing itself.

Once a new chemical passes this hurdle and production commences, the Environmental Protection Agency places it on the TSCA inventory. Thereafter it may be manufactured or used for most purposes by anyone without further notice to the agency. However, there are circumstances under which the EPA might decide not to regulate a new chemical because of limited exposure in the use originally intended, but would be concerned if certain other uses were intended entailing greater exposures. For example, if a potentially hazardous new chemical is initially used in a manufacturing process where exposure to workers is tightly controlled, the EPA might decide that such use would not pose an unreasonable risk. This situation could change, however, if the chemical were to be used in final consumer products—as a component of a spray paint, for instance. In such cases the EPA can promulgate what is known as a Significant New Use Rule for the chemical, thus subjecting specific uses of the chemical to the same reporting and review requirements as any new chemical.

Information collection. In addition to the testing and control powers TSCA gives the EPA, it contains several broad and useful provisions relating to information collection:

- The EPA can require submission of reports by manufacturers or processors concerning chemical production, use, disposal, environmental and health effects, and exposure.
- Manufacturers, processors, and distributors must keep records of significant adverse effects of chemical substances and mixtures. The specific circumstances of such recordkeeping are determined by the EPA through rulemaking.
- Manufacturers, processors, or distributors must report to the agency information that suggests a chemical substance or mixture presents a substantial risk to health or the environment. These reports provide the EPA with an early-warning mechanism to identify emerging concerns about chemicals.

Referral to other regulatory agencies. Although the Environmental Protection Agency seemingly has broad authority to regulate existing commercial chemicals, its scope of action may be severely circumscribed by another key section of TSCA. This provision apparently stems from Congress's recognition that activities under TSCA could overlap those taken under other statutes administered by the EPA as well as by other federal agencies. With respect to the latter, TSCA establishes a rather awkward coordinating mechanism. If the EPA determines that a chemical presents an unreasonable risk, *and* makes a finding that the risk can be reduced to a sufficient extent by action taken under a federal law administered by another agency, then the EPA *must* make a formal referral of the problem to the other agency. If the other agency either concludes that the risk is not as described by the EPA or initiates risk-reduction action, the EPA cannot take regulatory action. Only if the other agency finds that it does not have sufficient authority to address the risk or fails to initiate action can the EPA regulate. With respect to other laws administered by the EPA, there is somewhat greater discretion for the EPA to use TSCA when it is in the public interest to do so. Until now, most of the chemicals identified as candidates for regulation under TSCA have involved significant worker or consumer exposure, thereby potentially falling within the mandates of the Occupational Safety and Health Act or the Consumer Product Safety Act.

The wording in TSCA concerning referral is general enough to permit a variety of interpretations of the EPA's obligations to other agencies. For example, considerable controversy was generated when the EPA tentatively decided, then reconsidered the decision, to refer asbestos to the Occupational Safety and Health Administration and the Consumer Product Safety Commission for regulation. Central to the issue of referral is TSCA's intended role in regulating risks from existing chemicals. Is TSCA to be used solely as a gap-filling mechanism, germane only in the (relatively rare) cases where no other authorities are applicable? Or is TSCA to be used much more actively as a control on risk over the life-cycles of widely used chemicals? These issues have not yet been settled.

"Unreasonable risk" and the balance of benefits and costs. The concept of unreasonable risk is central to decision making under TSCA, but is never formally defined in the legislation. The legislative history of the act does provide some guidance, however:

> . . . In general, a determination that a risk associated with a chemical substance or mixture is unreasonable involves balancing the probability that harm will occur and the magnitude and severity of that harm against the effect of the proposed regulatory action on the availability to society of the benefits of the substance or mixture, taking into account the availability of substitutes for the substance or mixture which do not require regulation, and other adverse effects which such proposed action may have on society . . .
>
> . . . The committee [House Committee on Interstate and Foreign Commerce] has limited the [EPA] Administrator to taking action only against unreasonable risks because to do otherwise assumes that a risk free society is attainable, an assumption that the committee does not make.[22]

This interpretation is generally consistent with the notion (introduced in chapter 2) of balancing costs and benefits in deciding whether and how to regulate. (It should be noted that the House report from which this legislative history is taken specifically states that a quantitative cost-benefit analysis is not required to show unreasonable risk, although such an analysis is not precluded.)

The Federal Insecticide, Fungicide, and Rodenticide Act. Federal regulation of pesticides goes back to the turn of the century, when the rapid expansion of pesticide use began.[23] The purpose of the Insecticide Act of 1910, however, was to protect the consumer from fraudulent goods rather than to provide environmental protection. In 1947 Congress passed the Federal Insecticide, Fungicide, and Rodenticide Act, to be administered by the U.S. Department of Agriculture.[24] This act created a registration program for all pesticide products and established labeling requirements. As in the earlier legislation, the principal focus was on ensuring the effectiveness of pesticide products sold in commerce.

Beginning in the 1960s, concerns over the health and environmental effects of pesticides led to pressure for revisions to the pesticide statutes. Rachel Carson's book *Silent Spring* played a pivotal role in focusing these concerns. Amendments to FIFRA in 1964, 1972, and 1978 established the current framework for pesticide regulation, which has been administered by the Environmental Protection Agency since 1970.

The heart of FIFRA mandates that no pesticide can be introduced into commerce in the United States unless it has been registered by the EPA. The criterion for registration is that " . . . when used in accordance with widespread and commonly recognized practice [the pesticide] will not cause unreasonable adverse effects on the environment." Unreasonable adverse effects are defined as " . . . any unreasonable risk to man or environment, taking into account the economic, social, and environmental costs and benefits of the use of any pesticide."[25] Thus the basic decision-making criteria for TSCA and FIFRA appear to be quite similar. Both provide the EPA with broad latitude to balance costs and benefits in its decisions, in contrast to many of the other statutes administered by the EPA.

In implementing the pesticide registration program, the EPA can establish data requirements for its review and restrict the conditions of registration in various ways. FIFRA also recognized the problem associated with those pesticides registered long before current regulatory and scientific standards were established, and requires the EPA to reevaluate and reregister such pesticides. However, until amendments were enacted in 1988, FIFRA did not mandate a schedule for such review.[26]

Despite the requirement to review all existing pesticides, FIFRA, like TSCA, treats new and existing products somewhat differently. During registration of a new product, the burden is on the manufacturer to demonstrate the product's safety. In taking action to cancel or suspend an existing pesticide, on the other hand, the burden is shifted to the agency to demonstrate that an unreasonable risk exists. Under FIFRA this involves an elaborate and lengthy process, including a formal administrative hearing, a trial-like proceeding where both sides can present and cross-examine witnesses. Industry can then appeal the administrative decision to a judicial review.

In addition to the provisions of FIFRA, the EPA, under the Federal

Food, Drug, and Cosmetic Act, also establishes maximum permissible concentrations of pesticides in or on raw agricultural commodities and processed foods. Once established by the EPA, these limits are enforced by the Department of Agriculture and the Food and Drug Administration. In setting these standards, the EPA is subject to certain provisions of the Food, Drug, and Cosmetics Act, including the notorious Delaney clause which sets a zero-risk level for carcinogens in food additives. The inconsistency between this provision and the risk balancing approach in FIFRA has yet to be resolved.[27]

Comparison of TSCA and FIFRA. TSCA and FIFRA share important features. As we have seen, both regulate chemical products themselves rather than just the emissions that result from the production of those products, and both use similar risk-benefit criteria for decision making. Also, both provide for review of new chemicals before their commercial introduction, and provide mechanisms for addressing hazards associated with chemicals already in commerce. Both provide for a broad array of control measures, including total bans on the manufacture or use of a chemical, and both are concerned with exposure patterns across environmental media—air, surface water, and groundwater.

But the differences between the two statutes are at least as important as their similarities. The two laws are compared in table 6–3. In the case of new chemicals, TSCA requires only a limited premanufacture notification and imposes no up-front testing requirements. Ninety days after notifying the EPA, manufacture can commence unless the agency has taken some affirmative action to prevent it. In contrast, FIFRA contains a true registration program; a new pesticide cannot be introduced into commerce unless the EPA registers (approves) it. A manufacturer must show that the chemical poses no unreasonable risk by providing appropriate test data in advance. This seeming anomoly between the two laws is in fact quite consistent with the unreasonable risk concept that underlies both statutes. Pesticides are by their very nature intended to be biologically active materials. Therefore there is a greater probability of adverse consequences to humans and non-pest organisms from pesticides than from general commercial chemicals, thus justifying a higher threshold for approval.

These differences between TSCA and FIFRA are reflected in the costs borne by manufacturers under the two programs. According to one source, field and toxicology tests for a major new ingredient to be used in pesticides can take up to five years to complete and can cost from $5 to $7 million.[28] In contrast, the EPA has estimated that a firm wishing to introduce a new chemical into commerce incurs costs of only $1,300 to $7,500,[29] and the delay is limited to the 90-day review plus the time

Table 6–3. Comparison of TSCA and FIFRA Provisions

Category	TSCA	FIFRA
New products		
New chemical substances	Must be submitted for review 90 days before manufacture; manufacturer can commence production unless EPA acts within the review period	New active ingredients cannot be sold unless registered by EPA
New manufacturers of existing products	Not covered unless there is a Significant New Use Rule	Must receive registration
New product formulations	Not covered unless there is a Significant New Use Rule	Must receive registration
New use of product	Not covered unless there is a Significant New Use Rule	Must receive registration
Test data	No data required before submission	Extensive data required to register a new active ingredient
Exemptions	R&D, test marketing, others by rule if no unreasonable risk is found	Experimental use permits, emergency uses, and state registrations for special local needs allow for partial exemptions from the full registration process
Existing products		
Data gaps	Can require testing through rulemaking; EPA must show need for data	Can require data to reregister or maintain registration of existing pesticides; registration can be suspended if data are not provided
Control of unreasonable risks	Through rulemaking	Administrative proceeding to cancel or suspend registrations

needed to complete the form. Of course, if the EPA does take regulatory action on the chemical, additional costs and delays will result.

Under FIFRA, moreover, every new manufacturer of an existing pesticide, every new formulation utilizing existing active ingredients, and every new use of an existing product must be registered. In fact the vast majority of the 15,000 or so registrations processed each year by the EPA's Office of Pesticide Programs are for such situations; only about 15 to 25 significant new active ingredients are introduced each year, and there is a total of only 600 existing ingredients now under registration. On the other hand, manufacturers are required under TSCA to submit new formulations or uses only if the EPA has promulgated a Significant New Use Rule for the specific use and chemical. To date, relatively few such rules have been promulgated, and the Office of Toxic Substances (OTS) has received no notices of new use; it is currently reviewing more than 2,000 new chemical substances annually, however.

In the case of products already in use, FIFRA again gives the EPA more leverage than TSCA in gathering toxicity information. Basically, pesticide manufacturers must provide data required by the EPA in conducting its assessment of products. If a manufacturer chooses not to do so, it faces loss of the registration. Under TSCA, on the other hand, the EPA must go through the time-consuming process of issuing a regulation in order to gather test data.

ISSUES IN REGULATING CHEMICALS

Against this historical and legislative backdrop, it is important to ask several questions. Perhaps the most basic concerns the net accomplishments being generated through the current process of reviewing and regulating chemical products under TSCA and FIFRA. That is, how do the health and environmental benefits generated by these approaches to toxic chemical regulation compare with the social costs incurred? Are these statutes, and the EPA's implementation of them, yielding something close to the maximum in net benefits achievable in addressing the risk posed by commercial chemicals? As we have seen, decision making under both statutes must incorporate the notion of unreasonable risk, a rough cost-benefit criterion. Yet the mere ability to consider costs and benefits does not ensure that decisions will ultimately result that maximize net benefits to society. Moreover, other aspects of the legislation may make it difficult to achieve results consistent with cost-benefit-based decision making. For example, we have seen that both TSCA and FIFRA impose different regulatory burdens on new and existing chemicals. This differential treatment must be examined in light of the relative costs and benefits of regulatory activity in each area.

Unfortunately, it is not possible at present to answer these questions rigorously. This is true for a number of reasons. First, even under the best of circumstances it is difficult to evaluate quantitatively the risks associated with toxic chemicals, given the current state of knowledge. Second, both statutes, but TSCA in particular, provide regulatory tools to address information gaps in our knowledge of chemical toxicity and exposure. Ultimately, the benefits of such tools stem from the expected value of the information to be achieved, that is, its contribution to improved identification of chemicals of concern and their control. Although there has been some interesting research in this area, at present it is impractical to evaluate the net benefits of programs to test the toxicity of chemicals or gather other information on chemicals.

A third reason is that the time periods for implementing TSCA and, to a lesser degree, FIFRA have been short relative to the time periods associated with many of the effects of concern. Cancers associated with new chemicals currently being reviewed by the EPA, for example, might not be evidenced for ten or twenty years or more. Likewise, long-term environmental effects will be difficult to ascertain for many years.

Still another reason is that preventive laws such as TSCA and FIFRA are inherently difficult to evaluate because their most significant effects may never be seen. These effects are associated with toxic chemicals that never reach the stage of agency review, or chemical innovations that are deterred because of changes in company product-development decisions brought about by the requirements for agency review. Our ability to predict the nature of these changes is extremely limited.

Bearing these difficulties in mind, the discussion in this section has a somewhat more limited focus than that implied by the questions raised above. We focus instead on the progress that has been made and the problems that have arisen during implementation of the major objectives set forth in the statutes: obtaining adequate information on chemicals, controlling unreasonable risk from existing chemicals, and preventing future unreasonable risks from new substances while limiting adverse impacts on innovation. For each of these areas we attempt to provide some sense of what has been accomplished, the procedural and technical problems that have been encountered, and what has been learned about toxic chemicals regulatory policy. The focus is primarily on TSCA, although reference is also made to FIFRA where appropriate.

We now turn to the types of information that must be gathered to make determinations of unreasonable risk in toxic chemical regulation, as well as the problems encountered in gathering the information. This discussion provides a necessary backdrop for the subsequent three subsections on chemical information, existing chemical regulation, and new chemical regulation.

Toxic Chemical Risk Assessment

Toxic effects of the type that receive the most attention under TSCA and FIFRA are extremely difficult to assess. Adverse effects may not occur until long after the onset of exposure, and most people are exposed to complex combinations of chemicals rather than only the substance(s) under consideration. It is therefore difficult to establish causal relationships between human exposure to specific substances and the incidence of particular diseases. With a few notable exceptions, as in the case of asbestos, the determination of human health hazards must be assessed primarily on the basis of animal studies.

For example, the following steps are necessary to evaluate potential carcinogenic risks:

1. An experimental study (bioassay) is performed, utilizing small numbers of rodents. The chemical being tested is administered by one of several methods to groups of each sex for two species of rodents. Two or more dose-level groups are used, as well as zero-dose controls.[30]
2. At the termination of the experiment (usually the expected lifetime of the test species, about two years) the animals are sacrificed and their tissues examined to identify evidence of tumor formation.
3. The results are tabulated and their statistical significance assessed in combination with other available data on the substance.
4. If the data support a conclusion that the chemical is an animal carcinogen—and thus a potential human carcinogen—a quantitative dose-response model is fit to the data. The test animals are usually given high doses to safeguard against missing important effects in the small numbers of animals that can be used in such studies. Therefore dose-response models must be used to extrapolate to much lower levels, often orders of magnitude lower, than the doses actually employed in the experiments.
5. The dose-response model is adjusted to account for such factors as the difference between human and animal size, life expectancy, and exposure patterns to come up with a model that estimates human cancer probabilities at appropriate dose levels.

Each of these steps can be and has been subject to considerable scientific debate. Given our current understanding of the process by which cancer is initiated, for example, the basis for selecting any particular dose-response model is not strong. Yet different models that have been proposed can result in cancer predictions that differ by a factor of 100 or more when extrapolated to low doses.[31] As difficult as carcinogenic risk assessment is, there is at least a set of procedures which would allow some quantitative evaluation of the risk. For other possible adverse health effects—such as reproductive effects and neurotoxicity—methods for predicting human incidence rates from animal studies by quantitative methods are not yet available.[32]

Hazard assessment is only one element of the analysis that needs to be done to determine whether a substance poses unreasonable risk. Hazard must be combined with exposure to determine overall risk. As often as not, suitable exposure data are unavailable for the chemicals of concern. Monitoring to fill these information gaps is extremely expensive and may involve the development of specialized techniques to measure specific substances at very low levels of concentration.

Even where adequate data are available for the chemical of direct concern, the issue of relative risk must be addressed. Regulatory action on a specific chemical will create the incentive to substitute other substances. Substitution will occur even if the regulatory action falls short of a ban, since any added control costs will make the regulated chemical more expensive and hence shift demand toward its substitutes. Yet frequently the risk associated with these substitutes is less well understood than that of the chemical of primary concern. As a result, the *net* risk reduction associated with a regulatory option will be highly uncertain. Given all of these difficulties, it is clear that programs which seek to regulate toxic chemicals on the basis of the decision-making model described in chapter 2 do not have an easy task.

Information on Existing Chemicals

A driving concern leading to the enactment of TSCA was the belief that not only were certain chemicals hazardous, but that risks associated with the vast majority of them were virtually unknown because of the lack of adequate test data. The National Academy of Sciences (NAS) has provided some quantitative reinforcement for this belief.[33] As part of a study for the National Toxicology Program,[34] the NAS examined the adequacy of data available for performing risk assessment on chemicals. The procedure was to select a stratified sample of chemicals from different categories of use, then to utilize available data bases to determine what health effects data were available for each chemical.

The study drew from a universe of 65,725 chemicals consisting of those on the TSCA inventory, pesticides and inert ingredients, food additives, drugs, and cosmetic ingredients. From this universe, a stratified random sample of 675 chemicals was scrutinized to determine whether toxicity information was available. The results of this sample were then extrapolated back to the original universe.

Some of the results of this most useful analysis are presented in table 6–4. It is important to note that the search strategy used to identify toxicity data on the chemicals in the NAS sample was not exhaustive, and that additional data may be available in files of companies and government agencies. Nevertheless, for most chemical categories in the analysis the results are probably quite representative. These results suggest that significant data deficiencies exist, particularly for substances on the TSCA inventory. The latter category was subdivided into chemicals with aggregate production exceeding one million pounds in 1977, chemicals with aggregate production of less than one million pounds, and chemicals for which production data were unknown or inaccessible.

For example, data on possible chronic toxicity were available for only 4 percent of the chemicals on the TSCA inventory for which annual production exceeds one million pounds. In contrast, data on potential chronic toxicity were available for 39 percent of the drugs examined. Intuitively, one would have expected that more testing would have been done for the higher-volume chemicals. It is surprising that this was not done for the sample; toxicity data were lacking altogether for 78 percent of the high-volume category and 76 percent of the low-volume category. For an even greater percentage of the chemicals produced in unknown quantities there were no toxicity data.

Relatively speaking, more test information is available for drugs and pesticides than for commercial chemicals in general. The former are subject to registration programs requiring considerable up-front testing, unlike the TSCA review requirements, so this finding is not unexpected. Moreover, these programs have had more time to have an impact.

These results should be qualified in a number of ways. For example, chemicals on the TSCA inventory include a large number of polymeric materials. Although toxicity data may be unavailable for many such substances, there is strong reason to believe that due to the size and stability of many polymers they are unlikely to be biologically active. The situation is also somewhat less discouraging with respect to pesticides; for only 38 percent of the pesticides studied was information lacking in all of the data categories examined. Many of these "pesticides" may in fact be inert ingredients, such as solvents, which are not subject to registration as active pesticide ingredients.[35] Such considerations aside, however, the

Table 6–4. Availability of Toxicity Data on Chemicals

Category	Estimated percentage with prescribed test for:					% having no toxicity information
	Acute toxicity[a]	Subchronic toxicity[b]	Chronic toxicity[c]	Reproductive or developmental biology[d]	Mutagenicity[e]	
Pesticides and inert ingredients of pesticide formulations	59 (49–70)	51 (41–62)	23 (15–32)	34 (25–44)	28 (20–38)	38 (28–49)
Cosmetic ingredients	39 (32–47)	29 (23–36)	16 (11–21)	22 (16–28)	23 (17–30)	56 (49–64)
Drugs	75 (67–85)	62 (53–73)	39 (30–50)	45 (35–56)	32 (23–42)	25 (16–34)
Food additives	47 (40–55)	32 (27–41)	13 (9–19)	20 (14–26)	23 (18–30)	46 (39–54)
TSCA inventory						
> 1 million lb/yr	20 (15–25)	10 (7–14)	4 (3–7)	6 (3–9)	9 (6–13)	78 (73–84)
< 1 million lb/yr	22 (15–29)	8 (5–13)	3 (2–6)	4 (2–7)	10 (5–15)	76 (69–83)
Unknown or inaccessible	15 (9–21)	7 (3–11)	3 (0–6)	7 (3–12)	8 (4–13)	82 (76–89)

Note: Numbers in parentheses are 90-percent confidence intervals.

Source: National Academy of Sciences, *Toxicity Testing: Strategies to Determine Needs and Priorities* (Washington, D.C., 1984).

[a] Tests for acute toxicity via various routes of administration, as well as skin and eye irritation and skin sensitization.

[b] Repeated dose studies ranging from 14 days to 12 months.

[c] Lifetime chronic toxicity, including carcinogenicity.

[d] Tests for reproductive or developmental effects.

[e] Tests for the ability of a chemical to induce genetic changes, frequently used as screening tests to indicate potential for carcinogenicity.

results of the National Academy of Sciences study give credence to one of the principal concerns which led to TSCA and the revisions to FIFRA: that toxicity data are lacking on many, if not most, chemical products manufactured, processed, and used in the United States.

What resources would be required to improve significantly the amount of information available on chemical toxicity? Table 6–5 attempts to suggest an answer to this question, at least in part. It presents estimates of the costs of testing commercial chemicals at two different levels and for several thresholds of production volume. The less expensive tests ($250,000) would permit ascertainment of basic physical/chemical properties, possible acute health and environmental effects, potential mutagenicity (which might serve as a basis for predicting carcinogenicity), and subchronic toxicity, as well as reproductive and teratogenic effects. The more expensive tests ($1.25 million) would add a two-year bioassay for carcinogenic effects. For each of these test options, costs have been calculated assuming that all commercial chemicals above a certain annual production cutoff would be tested. For example, of the approximately 60,000 chemicals on the TSCA inventory in 1977, only 4,200 were produced in volumes above one million pounds. The costs of testing all such chemicals with the less expensive test battery, annualized over a 10-year period, would be $40 million per year.[36] These costs probably overestimate the true costs to some degree, since they have not been adjusted to account for data already available on chemicals or testing that may already be under way. On the other hand, costs associated with agency review of test results have been ignored, as have any costs associated with requiring testing through rulemaking.

The results indicate that any comprehensive testing program would require substantial time and resources. But these costs are not of themselves enough to discourage broad testing of chemicals. Even the most expensive option considered—level II testing of all chemicals produced in volumes greater than one million pounds—results in annualized costs equal to about 0.3 percent of annual industrial chemical sales.

It is important to note, however, that even if such an ambitious testing program were initiated, it would not guarantee the availability of data suitable for a risk assessment for many of the chemicals tested. As pointed out previously, a risk assessment requires information on both toxicity (or hazard) and exposure. Yet exposure data are frequently inadequate. Thus a balanced program of data-gathering would have to include exposure assessments as well as toxicity testing.

Activity under TSCA. As we have seen, TSCA provides a regulatory mechanism for obtaining test data on chemicals in commerce. How much progress is the EPA making in identifying information needs and getting

Table 6–5. Cost Estimates of Testing Programs to Improve Information

Testing level	Production volume cutoff (10^6 lb)	No. of chemicals	Years to complete[a]	Annualized costs (\$$10^6$)[b]	Annualized costs as a percentage of chemical industry sales[c]
Level I	1	4,200	42	40	0.07
(total cost = \$250,000)[d]	10	2,000	20	35	0.06
	100	800	8	22	0.04
Level II	1	4,200	42	200	0.33
(total cost = \$1,250,000)[e]	10	2,000	20	173	0.28
	100	800	8	108	0.18

Source: R. C. Evans, J. Bakst, and M. Dreyfus, "Analysis of TSCA Reauthorization Proposals," draft report, Office of Pesticides and Toxic Substances, Environmental Protection Agency, 1985.

[a] Assumes a testing capacity of 100 new studies per year.

[b] Equivalent annual costs required to pay for the testing program over a ten-year period, computed at a 10 percent discount rate. Costs have been annualized to facilitate comparison across programs that would take different lengths of time to complete.

[c] Annual costs divided by 1983 annual sales of industrial chemicals, \$60 billion. This is a subset of total sales of chemicals and allied products.

[d] A hypothetical test scheme including testing for basic chemical and physical properties, acute toxicity, subchronic toxicity (90-day study), mutagenicity, reproductive and teratogenic effects, and ecotoxicity.

[e] Level I plus additional testing, including a two-year cancer bioassay.

tests under way? The Environmental Protection Agency has focused most of its efforts to date on addressing those chemicals nominated by the Interagency Testing Committee. This is natural, since the agency is required by law to either initiate or decline action within one year of designation by the committee.

The EPA had considerable difficulty meeting the one-year deadline on the committee's early designations. This failure led to a lawsuit, which in turn resulted in a court-ordered schedule on which these chemicals must be tested.[37] Since the lawsuit, the EPA has met the statutory deadlines for those chemicals covered by the court order and has met the statutory deadlines on all subsequent ITC designations. In 1985 the EPA initiated for the first time test rules for chemicals other than those referred by the ITC.[38]

Yet in light of the apparent information needs, progress in getting chemicals tested appears to have been excruciatingly slow. Through the end of fiscal year 1986, final test rules had been issued for 13 chemicals or chemical categories. Negotiated testing agreements were in place to obtain voluntary testing of an equal number.[39] The EPA decided not to issue test rules for 45 chemicals or categories because available data were adequate for regulatory decision making or because ongoing testing by industry or government was expected to eliminate existing data gaps.

It should be emphasized that section 4 of TSCA is not the only mechanism used to get test data on commercial chemicals. The government does a certain amount of toxicity testing, and voluntary testing by individual firms and by industry-funded institutes contributes significantly to the knowledge base. Nevertheless, progress under TSCA has been disappointing in view of the magnitude of the problem described by the National Academy of Sciences. Why has there not been more activity since 1977? Many factors have probably contributed, including the EPA's difficulty in figuring out how to implement this novel regulatory program. In particular, the agency was probably overly cautious with respect to the level of analysis required to support testing decisions. Nevertheless, the most important determinant of the slow pace of testing has probably been TSCA itself.

The act requires that a new regulation be written each time it is desired that a chemical be tested. This is inherently time-consuming. Moreover, each time the EPA wishes to regulate on the basis of unreasonable risk, it must muster comprehensive and very detailed evidence in support of its actions. When applied to individual chemicals, these requirements severely restrict the number of substances that can be addressed. In some cases the toxicity testing ultimately required under TSCA is less costly than the rulemaking process the EPA must use.[40] Despite continuing efforts to improve the process, it appears that the principal route for expediting and expanding chemical testing is to make use of broad-based testing approaches, either through more aggressive use of the current statute or through modifications to the statute, as some have already proposed.[41]

In considering strategies to improve the *quantity* of testing (or exposure) data, however, we should not lose sight of an important point. Test data are not an end in themselves; they are useful only if they facilitate improved decisions on the management of chemical risks. Testing data in the absence of a plan to use them may not be cost-effective in meeting TSCA's objective that "adequate" data be available for assessment. Viewed from this perspective, the timing of chemical tests for substitute chemicals is an interesting issue.

Regulatory action on individual chemicals must consider the nature of available substitutes, including efficacy and price, for this will determine the initial costs and ultimate economic impacts of regulating a particular chemical. Equally important, the risks associated with these substitutes will determine the *net* risk reduction associated with the regulatory action. As we have seen, TSCA requires that the EPA take into account these aspects of chemical substitution. Moreover, the characteristics of substitutes can have an important impact on regulatory strategies. For example, if all chemicals that could be used in a particular application offer roughly comparable levels of risk, then general strategies aimed at reducing exposure to the whole cluster of chemicals may be appropriate. Where there is a considerable difference in risk among substitutable chemicals, however, bans on the most hazardous of them might be a preferred strategy.

In contrast to these considerations, under TSCA chemicals have been singled out for test rules mostly on a one-at-a-time basis, with occasional exceptions. Thus under the act there is a real danger that even after the long process of testing and data evaluation is complete, adequate data for regulatory decisions will not be available because of the lack of suitable data on substitutes. This is a potentially serious problem under the EPA's implementation of the law.

Under FIFRA, the EPA has addressed this same problem and has developed a solution that involves collecting pesticide reregistration data for chemical-use clusters, consisting of groups of chemicals used for similar applications. The benefit of this approach is that data can be requested and ultimately obtained contemporaneously for chemicals that are likely to substitute for each other. A similar concept should be embraced for TSCA chemicals, although the latter encompass a much wider spectrum of use.

Controlling Unreasonable Risk from Existing Chemicals

If the EPA finds that an existing chemical poses an unreasonable risk to human health or the environment, the agency can regulate it under TSCA or refer it to another agency or another office of EPA for regulatory consideration, as we have seen. Since 1977, however, neither has occurred very often. As of 1986, the EPA had elected to regulate only one group of chemicals under the provisions relating to unreasonable risk; the agency banned the use of chlorofluorocarbons as aerosol propellants because of their potential for stratospheric ozone depletion. (The manufacture, processing, distribution, and use of PCBs were banned directly by Congress when it wrote TSCA; the EPA played no role in that decision.) Two specific chemicals, methylene dianiline (MDA) and butadiene, and a group of four solvents have been formally referred to the Occupational Safety and Health Administration for regulatory consideration.[42]

Why so little regulatory activity over the past decade? Part of the reason has to do with program priorities. Much of the early effort in implementing TSCA was devoted to organizing and staffing the new office, meeting statutory deadlines for promulgating test rules, compiling the TSCA inventory, and putting the new chemical review program into place. These and other activities were time-consuming.

At least as important has been the basic difficulty of making unreasonable risk determinations in the face of the data limitations and uncertainties discussed above. Efforts to deal with the chemical formaldehyde illustrate some of the difficulties involved in attempting to manage risks. Among chemicals in commerce, formaldehyde is one of those produced in the largest volume; about 5.8 billion pounds of the solution were produced in 1985, and the solution ranked 24th in total production among all chemicals in commerce.[43] Although some formaldehyde solution finds direct use (in tissue preservation, for example), its major use is as an intermediate in the production of plastics, adhesives, and permanent-press fabrics. Formaldehyde-based adhesives are widely used in the manufacture of particle board and plywood. Although most of the formaldehyde is consumed in the production of these materials, some "free" formaldehyde often remains. Moreover, certain chemical reactions can degrade the materials, releasing free formaldehyde during use. Exposure to formaldehyde vapors is therefore extensive, so any associated toxic effects are of significant concern.

Formaldehyde has been known to be an irritant to the eyes, nose, and throat at concentrations as low as 0.1 ppm in the air. At much higher levels it is acutely toxic. But in 1979 the Office of Toxic Substances first received interim bioassay information from an industry-sponsored testing lab indicating that formaldehyde was carcinogenic in laboratory animals. In the following year a federal panel, convened to review the preliminary results of the full study, found that the bioassay was consistent with accepted testing standards and that the chemical should be presumed to pose a risk of cancer in humans. The new bioassay data led the EPA to consider whether formaldehyde presented a "significant risk of serious or widespread harm to human beings from cancer, gene mutations, or birth defects"—the criterion in TSCA for quick action. If formaldehyde met this criterion, the agency would have to initiate regulatory action within 180 days or find that the risk was not unreasonable.

In February 1982, nearly three years after receiving first word of the animal study, the EPA announced that formaldehyde did not meet the criterion for expedited regulatory action. In doing so, the agency acknowledged that formaldehyde was a potential human carcinogen, but argued that the available data on the degree of carcinogenic potential, combined with exposure, did not justify the high priority afforded an accelerated rulemaking. Given the strength of the bioassay findings and the widespread use of the chemical, this decision created considerable controversy. Indeed, the Natural Resources Defense Council and the American Public Health Association sued the agency over this decision.

Before the issue could be decided in court, the Environmental Protection Agency reopened its decision, asking for public comment on the issue in November 1983.[44] The following May, the EPA announced that certain uses of formaldehyde did in fact meet the conditions for accelerated regulatory consideration. These were formaldehyde used in the textile industry, where 777,000 workers were estimated to be exposed to the chemical from permanent-press fabrics, and in home products, through which more than 100 million people may be exposed to formaldehyde from adhesives in such wood products as plywood and particle board. A large number of other formaldehyde uses were determined not to require expedited review, either because concentrations were low or because few people were at risk. Since 1984 the EPA has been struggling to complete its regulatory decision-making on formaldehyde. In May 1986 the agency referred the textile usage to OSHA, which regulated worker exposure to formaldehyde in textile and other workplaces by setting a maximum workplace exposure level. This leaves the EPA with the wood products usage. Will the EPA find that this use is an unreasonable risk under TSCA?

Such a finding is not a foregone conclusion because the costs of significantly reducing formaldehyde levels could be substantial. (It should be recalled that TSCA mandates a balancing of benefits and costs.) For example, the annual cost of replacing current adhesives with formaldehyde-free substitutes could be $70 million just for wood products going into new home construction.[45] However, much less expensive options are also available, involving use of materials that substantially reduce, but do not eliminate, formaldehyde emissions. Much additional work has to be undertaken to understand the complex relationships that determine formaldehyde release from products and the relationships between formaldehyde release and actual exposure levels in the home or workplace.

To complete the circle, even the original toxicity data on formaldehyde, while strong, leave considerable room for interpretation. Formaldehyde did test positive in at least one species of animal in a well-conducted bioassay. And epidemiological studies provide some limited evidence that the chemical may be carcinogenic in humans. Moreover, close analogs of formaldehyde are also animal carcinogens, and formaldehyde has tested positive in certain short-term screening studies that test for the potential to cause cell mutations. Nevertheless, considerable uncertainty remains over the significance of the cancer risk posed to *human* populations. For example, the data linking dose and response can be interpreted in various ways, using different mathematical procedures to extrapolate from the high exposures where effects were observed in animals to the generally lower levels to which human populations are exposed. Depending on the technique, the EPA estimates of individual cancer risk from exposure to formaldehyde in homes varied by as much as 1,000 times.[46]

If the EPA makes an initial determination that formaldehyde risks are unreasonable for wood product uses, it must then take into account what TSCA has to say about referral to other agencies. The Consumer Product Safety Commission and the Department of Housing and Urban Development (HUD) arguably can regulate much of the residential exposure. Indeed, HUD has already acted to limit the release of formaldehyde from building materials used in manufactured homes, although it did so on the basis of formaldehyde's irritating rather than potentially carcinogenic properties.

Formaldehyde provides an excellent example of the difficulties involved when a major existing chemical product is being considered for regulation on the basis of a chronic toxic effect such as cancer. With the economic stakes as high as they are for a chemical like formaldehyde, the many uncertainties associated with the assessment of chronic hazards are subject to intense scrutiny and debate. Thus nine years after the first preliminary toxicity data were made available to the EPA, the agency still had not completed its rulings on one of the chemical's highest priority uses.

Although the procedures are somewhat different, many of the same problems arise in regulating economically important pesticides. It took the EPA about seven years to ban ethylene dibromide, a fungicide used to fumigate grain.[47] Of course, the data on which decisions were being based changed drastically over this period. Thus the final resolution of this action was based to a significant extent on information that was not available at the start of the regulatory process. More generally, the EPA's special review process for regulating existing pesticides has taken from two to six years to complete, and can be followed by administrative hearings which could take another two years. Nevertheless, in contrast to the few actions taken under TSCA, under FIFRA the Environmental Protection Agency completed thirty-two special reviews of pesticides between 1975 and 1985.

Regulating New Chemicals: The Search for Balance

When the Toxic Substances Control Act was being developed, no issue received more attention than the procedures that should be used in considering new chemical substances. The controversy stemmed from the difficult problem of identifying and controlling risk at early stages in the life-cycle of a new chemical, while at the same time avoiding significant adverse impacts on beneficial innovation.

The basic requirements for premanufacture notification of new chemicals were contained in the 1976 law. Nevertheless, it took the EPA a long time to promulgate final regulations formalizing the reporting requirements; although originally proposed in January 1979, the EPA did not finalize these requirements for submitting PMNs until October 1983.[48] Even then, portions of the final rule were stayed by a court pending resolution of several issues.[49]

Despite the delay, however, the submission of premanufacture notices began in 1979 under interim guidelines. Annual data on the number of submissions are presented in figure 6–2. As the graph demonstrates, the number of notifications has increased substantially over time; currently more than 2,000 notifications are being submitted annually. The rapid growth in submissions probably reflects industry response to a regulatory transition rather than an underlying trend in new product development. Some firms, for example, may have accelerated new product introductions before 1979 in order to get their chemicals in the initial TSCA inventory, thereby avoiding PMN requirements. These firms would then have reduced their need to introduce new products for some time. Other firms may have held back new products until the EPA's procedures and policies for reviewing PMNs were demonstrated through experience with a few early examples.

About three-quarters of the chemicals for which premanufacture notice is received are of low concern and receive only a limited, initial review.

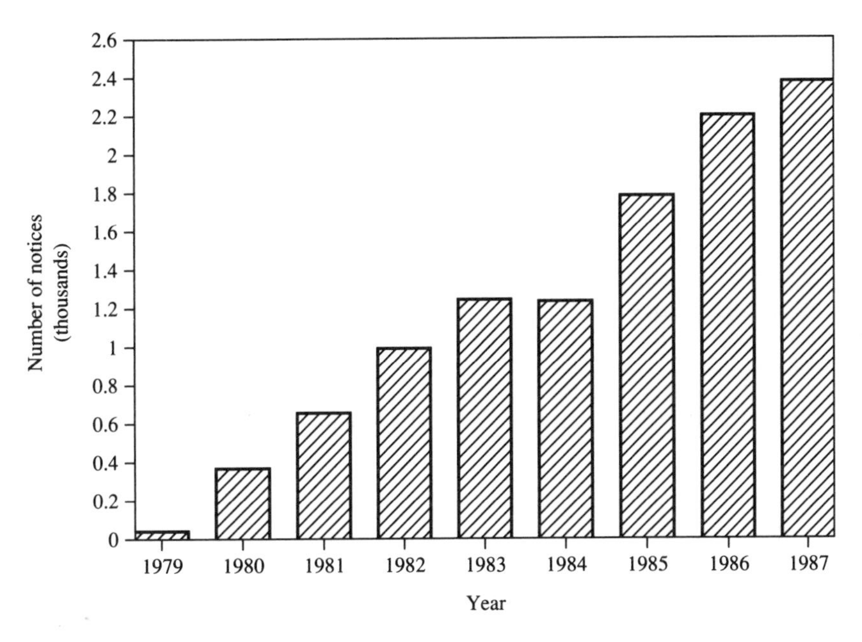

Figure 6–2. Submissions of premanufacture notices. Numbers for 1985–1987 include limited notifications submitted for polymers and low-production volume chemicals under regulations that became effective in 1985.

Thus manufacturers can commence production once the 90-day notification period expires. Of the chemicals that undergo more detailed assessment, many are ultimately dropped from review, but a significant number have been the subject of some form of regulatory action. Table 6–6 summarizes actions on premanufacture notifications submitted during the fiscal year 1984. Of 1,192 chemicals submitted, about 86 percent were ultimately dropped from review without action due to lack of supportable toxicity concerns and/or exposure. Roughly 4 percent of the chemicals were withdrawn by the companies wishing to market them because of concerns raised during the review, and 7 percent were the subject of formal regulatory orders. Another 2 percent were in suspended status as of 1986, awaiting the resolution of issues that arose during review. Some portion of these will ultimately be withdrawn or controlled.

The statistics cited above are for 1984 only. Since 1976 more than 10,000 new chemicals have been reviewed by the EPA under TSCA. Based on this experience, it is possible to reach some general conclusions about the PMN process and the nature of the issues that have arisen.

Table 6–6. Premanufacture Notice (PMN) Dispositions, 1984

Action taken	No. of PMNs
Dropped from further review by the EPA[a]	1,029
Withdrawn by the submitter	53
Regulatory action[b]	83
Suspension[c]	27
Total	1,192

[a] Chemicals dropped from PMN review without any EPA action. About 15 percent of these cases were referred internally for consideration as Significant New Use Rule candidates.

[b] Orders regulating manufacturing, processing, distribution, use, or disposal.

[c] Review period voluntarily suspended by the submitter in order to resolve questions raised by the EPA (as of January 1986).

The nature of new chemical introductions. The great majority of the so-called new chemicals reviewed by the EPA are not uniquely so. Rather, they represent relatively small modifications to existing substances. These modifications might enhance certain desirable characteristics (or diminish undesirable ones), thereby enhancing the attractiveness of the new substances relative to existing products. This is not to say, however, that the benefits to society from such new products are insignificant. Over time this process of evolutionary product development can yield substantial cumulative benefits, although the incremental nature of most new chemical innovation does present some interesting consequences for the EPA's review process, as described below.

Availability of test data. Comparatively few health or environmental data have been provided with most PMNs. One study found that 47 percent of the PMNs submitted through June 1983 included *no* health or environmental toxicity data, while only 11 percent had reported short-term animal toxicity studies and another 17 percent reported short-term mutagenicity test data.[50] Nevertheless, this record is somewhat better than the figures for all commercial chemicals reported in the National Academy of Sciences study discussed above. It should be kept in mind, also, that even without direct test data, the EPA is able to identify potentially toxic chemicals by examining other structurally similar chemicals for which test data are available. Indeed, such structure-activity analysis is a key component in the review of new chemical introductions. It goes without saying, however, that more and better data are desirable.

The Environmental Protection Agency does have the option of requiring more data on proposed new chemicals, but economic realities limit the potential for generating test data through this mechanism. Because many new chemicals represent incremental improvements over existing products, and are targeted for narrow markets, their ability to bear testing costs at the initial stages of commercialization may be quite limited. The costs and relative impacts of hypothetical testing options on a sample of chemicals that the EPA had reviewed through the end of 1984 are illustrated in table 6–7. The numbers in the table are calculated from estimates of test costs and the present value of life-cycle sales. The table presents the percentage of chemicals for which test costs would amount to 0 to 10 percent of total sales, 10 to 25 percent, and more than 25 percent. Three different levels of testing are considered—$10,000, $25,000, and $70,000. The results are presented for all chemicals in the sample and those for which annual production exceeded 10,000 kilograms per year.

As table 6–7 illustrates, the impacts of advance testing requirements could be significant. For instance, $70,000 in test costs (which would not go very far) would exceed 10 percent of the expected life-cycle revenues

Table 6–7. Costs and Impacts of Premanufacture Testing Requirements

Estimated cost of testing requirement	Exclusion[a]	Test cost as a percentage of present value of sales[b]		
		<10	10–25	>25
		Percentage of new chemicals falling within range		
$10,000[c]	—	85	5	10
$10,000[c]	<10,000 kg	99	1	0
$25,000[d]	—	77	8	15
$25,000[d]	<10,000 kg	87	9	4
$70,000[e]	—	60	15	25
$70,000[e]	<10,000 kg	82	12	6

Source: R. C. Evans, J. Bakst, and M. Dreyfus, "Analysis of TSCA Reauthorization Proposals," draft report, Office of Pesticides and Toxic Substances, Environmental Protection Agency, 1985.

[a] Excluding chemicals with less than the indicated annual production.

[b] Calculation based on a ten-year life-cycle projected from submitter's estimated production volume. Prices were estimated from prices of chemicals in similar uses. A 10 percent discount rate was assumed in calculating the present value.

[c] Minimum battery of acute toxicity tests and mutagenicity screening test.

[d] More comprehensive battery of acute tests and mutagenicity screening.

[e] Acute tests, subchronic tests, mutagenicity screen and ecotoxicity tests.

for 40 percent of the chemicals analyzed. (The average pre-tax profits for the chemical industry are less than 10 percent of revenues.)[51] New chemicals are frequently specialty products that can command higher profits than the industry average, but even at a 25 percent profit rate, a quarter of the PMN sample would incur costs equalling or exceeding this total profit level. Of course not all test costs would ultimately be absorbed by the companies producing the chemicals. Depending on the markets, some costs might be passed on to buyers, reducing the net impact.[52] The table does illustrate, however, that as applied to the universe of chemicals for which PMNs must be submitted, testing requirements can have a significant impact on the likelihood of new chemical introduction. More definitive tests, such as two-year bioassays costing about $800,000, would eliminate all but the most profitable new chemical introductions. The point is not that testing is undesirable, but rather that significant up-front testing requirements could frequently result in withdrawal of the PMN—that is, in a decision not to manufacture the chemical. In such instances society gets neither the test data nor the benefits of the new substance. Where there is a good basis for concern, such an outcome is quite appropriate. On the other hand, where the basis for concern is weak, the EPA is hard pressed to justify a de facto ban on a new chemical.

Consistency in Regulating New and Existing Chemicals

We have seen that TSCA treats new chemicals differently than existing ones. Manufacturers must submit proposed new chemicals to the EPA for review prior to manufacture and must await completion of the review before they can begin production. The EPA can use relatively simple administrative orders to prevent chemicals from being manufactured until needed test data are made available. On the other hand, a more complex (and lengthy) process is required to initiate testing or regulation of existing chemicals. Moreover, the latter can continue to be produced while rulemaking and testing are under way. This differential treatment of new and existing sources of risk reflects in part Congress's belief that preventive approaches are more effective than remedial actions. According to the legislative history of TSCA,

> . . . the most desirable time to determine the health and environmental effects of a substance, and to take action to protect against any potential adverse effects, occurs before commercial production begins. Not only is human and environmental harm avoided or alleviated, but the cost of any regulatory action in terms of loss of jobs and capital investment is minimized.[53]

Critics have argued, however, that in the long run a bias against new chemicals can result in both higher costs *and* increased risks: higher costs because economic progress is delayed; increased risks because the new products that would replace existing ones may be inherently less risky than the latter.[54] At least one critic of TSCA has argued that this is precisely what is happening as a result of the EPA's review of new chemicals,[55] and there appears to be some substance to the charge.

The evidence takes the form of comparisons of similar chemicals, some of which are new, others of which have been in commerce for some time. Since most new chemical regulatory actions are based on data for analogous chemicals, and since most new chemicals are themselves modifications of existing ones often used for the same purpose as the new chemical, it is not surprising that similar toxicity concerns would exist for both. Two of the most frequently regulated classes of new chemicals have been dyes and acrylates, which provide case studies on possible bias.

Dyes were the first commercially important synthetic chemicals. Despite the diversity of dyes which have evolved for textiles, leather, paper, and other products, most dyes are based on a relatively small number of basic chemical structures. A number of dyes, or intermediate chemicals used to make them, have been shown to be carcinogenic in humans or animals. Dyes sharing structural features similar to these materials are therefore suspect, and have been subject to regulatory action under the new chemicals program. The concerns over these materials usually relate to workers exposed in handling the dyes during manufacturing and application processes, although in some cases there is also concern about exposure of the general population through consumption of drinking water containing residual dye materials from waste-water effluents. When such concerns arise in new chemical review, the regulatory response has generally been to require testing, in some cases a two-year bioassay. In the latter instances, the chemicals have been withdrawn by the manufacturers, since most dyes have a limited commercial potential. However, no dyes in commercial production prior to TSCA have been banned or substantially regulated under that statute, even though they can be presumed to pose similar risks.

Acrylates provide another example. First introduced in the 1930s, more than a billion pounds of these materials are currently being produced. Acrylates have been associated with a number of adverse health effects, including skin and respiratory irritation and sensitization, and several have tested positive in carcinogenicity bioassays. Moreover, some applications in coatings, inks, and adhesives present the possibility of worker exposure to reactive acrylate materials. Ironically, such applications have been promoted in part because of their ability to reduce or eliminate the need for solvents, many of which are controlled under air pollution regulations, and to reduce the need for energy use in drying coatings.[56]

Since the mechanism by which acrylates cause cancer in laboratory animals is not well understood, proposed new varieties have been treated as potential carcinogens. Testing to resolve concerns would consist of full bioassay studies, the expense of which would preclude new acrylates from entering the market. In light of the potential benefits cited above, and the availability of control measures in most of these cases, the EPA has attempted to negotiate consent orders with those wishing to begin production—agreements which would permit manufacture but require protective clothing and, in some cases, respiratory protection for workers engaged in the manufacturing, processing, and use of the new chemicals or products containing them. In essence, these consent orders constitute the equivalent of control regulations that would require lengthy and expensive rulemaking if the EPA attempted to apply them to existing chemicals. Because of the lower threshold for EPA action on new chemicals, however, these requirements can be implemented easily for proposed new acrylates.

For both dyes and acrylates, the new chemicals being regulated represent small additions to existing commercial classes of chemicals, upon which the toxicity concerns are actually based. In many, if not most cases, the new chemicals will actually be substituting for existing members of these classes. Comparative toxicity assessments of these new chemicals have been hampered by a lack of confirmed, short-term predictive tests for many of the materials. Consequently, regulatory options have been limited to the imposition of what would prove to be unaffordable test requirements or negotiated exposure controls. In neither case will our knowledge of the toxicity of these chemical families be increased.

In contrast, the Environmental Protection Agency has required little testing or control of existing members of the chemical families under TSCA. Yet to the degree that additional test data are necessary to characterize these classes of chemicals, such testing is more likely to be affordable for existing commercially successful materials. Moreover, to the degree that these materials do present unreasonable risks, the majority of exposures will be to the existing materials. Thus a regulatory focus solely on the new substances is not cost-effective as a mechanism for mitigating risk from these classes of materials. Of course, such materials may or may not actually pose an unreasonable risk; the point is that the handling of new (as distinguished from existing) members of these classes has not been done in a manner that would increase the data base upon which such determinations can be made or that would address the widest exposures. Recognizing these limitations, the EPA began in 1985 to reassess its approach to these and other chemical classes, with the aim of better integrating approaches to dealing with both new and existing members of the categories.[57] However, to date there have been no significant results from this initiative.

The examples discussed are based upon chemicals that have actually undergone new chemical review by the EPA. As mentioned earlier, in the long run the major benefits and costs associated with TSCA are likely to be a result of decisions by firms to forgo development of new hazardous chemicals and/or beneficial compounds because of the regulatory review program. Much of the debate over the nature of the new-chemical review program and the requirements for up-front testing can really only be addressed in the context of these potential outcomes. Yet after nine years of experience with the new-chemical review program, relatively little can be said about the long-run changes in new chemical development likely to result from the TSCA program.

Some limited information is available on the number of new chemical introductions. Annual aggregate new chemical introductions by a sample of chemical firms from 1973 through 1984 are presented in figure 6–3. The numbers must be interpreted carefully, since different data sources were used for pre– and post–TSCA reporting periods. Qualitatively, however, the graph suggests that after an initial transition period, new

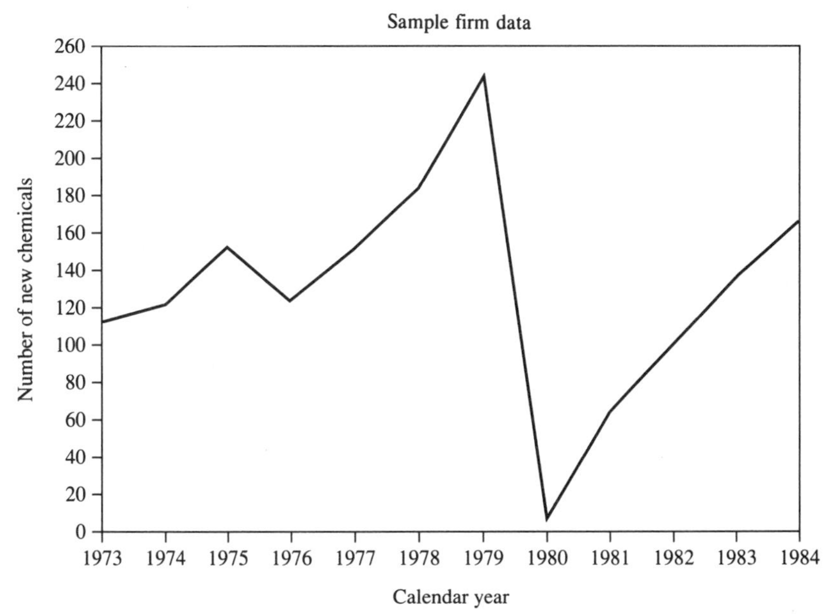

Figure 6–3. New chemical introductions, 1973–1984

chemical introductions have returned to a level comparable to that of the mid-seventies—that is, prior to TSCA. In contrast, a survey by the Chemical Specialty Manufacturers Association suggests that there has been some reduction in new chemical introductions, especially by smaller firms.[58] Both that study and the data in figure 6–3 are based on non-random samples and cannot be considered representative of the industry in general. Moreover, neither sheds any light on the economic value of new chemical introductions, which is more significant than the number of chemicals per se.

If information on the cost of forgone innovation is scanty, data on the change, if any, in the toxicity of new chemicals as a result of the PMN program are nonexistent. Thus it is not possible to conduct anything like a realistic cost-benefit analysis of the current program, let alone alternatives to it.

CONCLUSIONS

Despite considerable public concern about the dangers of exposure to toxic chemicals, for the majority of chemicals in commerce the available data are grossly inadequate to characterize potential chronic hazards, let alone to determine whether risks are unreasonable and permit selection of appropriate regulatory strategies. While a major objective of the Toxic Substances Control Act is the development of such data, almost ten years after its passage progress in reducing these data gaps has been quite slow. This has been so for two reasons. First, the EPA's focus on individual chemicals and narrow categories, coupled with the expensive and time-consuming regulatory process Congress mandated, has greatly limited the number of existing chemicals that have been tested. Second, the EPA's latitude for requiring tests of new chemicals is sharply narrowed by the generally limited commercial potential of most chemicals. In many cases the practical implication of testing requirements is a ban on the new chemical.

To a degree, these difficulties stem from the statutory framework that underlies TSCA. But it is not necessary to change the law to increase the availability of test data; even within the current statute there appear to be opportunities for broader information collection. Nor is it necessarily desirable to impose uniform testing requirements on all new or existing chemicals, as some might propose. Instead, testing and assessment strategies should be developed for broad categories of related chemicals. These strategies should be designed in a manner that can provide the targeted information needed to economically resolve regulatory issues concerning members of the specific chemical category. For example:

- Testing strategies should be designed around structurally similar classes of chemicals in a manner that would allow assessors to validate structure/activity relationships and short-term tests with a limited number of expensive long-term tests on members of the category.
- For some categories of chemicals, data on individual exposure rather than toxicity would most effectively serve the needs of decision making. While controversial, the EPA's ability to gather such data under TSCA should be pursued.
- In some cases, use-based categories should be defined, with chemicals that substitute for each other in use being tested in parallel. This would ensure that adequate information for relative assessment would be available.

These approaches will not resolve all of the problems discussed above. While rare, unique commercial chemicals do appear in the new-chemical review process, and the products of biotechnology are beginning to enter the review process. Category-based testing obviously cannot be applied to these types of materials. In such circumstances, the Environmental Protection Agency will continue to face difficult tradeoffs between the need to adequately assess and control risk and the desire to avoid delays in the introduction of beneficial technologies.

Even where data do exist, the regulation of existing chemicals will always be difficult because of the inevitable uncertainties in characterizing overall risks and the high economic stakes associated with commercially important substances. The current statutory framework, however, adds to rather than simplifies these problems. At first blush, TSCA appears to approach the regulation of toxic substances comprehensively, viewing risks from the perspective of the entire chemical life-cycle. In reality, however, TSCA is overlaid on a patchwork of other environmental laws and regulations which address the same risks from the standpoint of specific environmental media or exposure groups. Although interpretations differ on this matter, TSCA appears to imply that priority be given to these other laws when overlaps exist. As a result, the Toxic Substances Control Act has thus far contributed to further duplication of risk management investigations in the federal government.

This problem is probably best addressed through revised legislation. Absent such action, regulation of existing chemicals through TSCA is likely to be limited to a few cases where other authorities are inappropriate or inadequate. On the other hand, as we have seen, the task of obtaining adequate knowledge concerning the risks of existing chemicals is large enough that priority ought to be afforded to this activity in any event.

NOTES

1. Environ Corporation, "Principles of Risk Assessment: A Nontechnical Review," paper presented at a U.S. Environmental Protection Agency Workshop on Risk Assessment, Easton, Md., March 17–18, 1985.

2. R. Doll and R. Peto, "Avoidable Risks of Cancer in the U.S.," *Journal of the National Cancer Institute* vol. 66 (June 1981) pp. 1191–1308.

3. Ibid., p. 1256.

4. Toxic Substances Strategy Committee, *Toxic Chemicals and Public Protection* (Washington, D.C., Government Printing Office, 1980).

5. B. G. Reuben and M. L. Burstall, *The Chemical Economy* (London, Longman Group Limited, 1973).

6. Chemicals can be classified as organic or inorganic. Organic chemicals are compounds of carbon, while inorganic chemicals generally do not contain carbon. This nomenclature has its roots in the nineteenth century, when chemicals came either from living organisms (organic) or from minerals (inorganic).

7. "Facts and Figures for the Chemical Industry," *Chemical & Engineering News* vol. 65 (June 8, 1987) pp. 24–76.

8. Michael J. Bennett and Charles H. Kline, "Chemicals: An Industry Sheds Its Smokestack Image," *Technology Review* vol. 90 (July 1987) pp. 37–45.

9. G. K. Kohn, "The Pesticide Industry," in J. A. Kent, ed., *Riegel's Handbook of Industrial Chemistry* (New York, Van Nostrand Reinhold, 1983) pp. 747–785.

10. Ibid.

11. Council on Environmental Quality (CEQ), *Environmental Quality—1979* (Washington, D.C., Government Printing Office, 1979); Council on Environmental Quality, *Environmental Quality—1975* (Washington, D.C., Government Printing Office, 1975).

12. A chemical intermediate is a chemical used in the manufacture of another chemical product. In the case of vinyl chloride, that product is the widely used polymer, polyvinyl chloride (PVC). It should be noted that in many cases the properties of the end product differ significantly from the intermediate chemicals used to make it. Thus the health concerns for vinyl chloride do not extend to the PVC product.

13. Council on Environmental Quality, *Environmental Quality—1976* (Washington, D.C., Government Printing Office, 1976).

14. Toxic Substances Strategy Committee, *Toxic Chemicals and Public Protection.*

15. CEQ, *Environmental Quality—1976.*

16. Toxic Substances Strategy Committee, *Toxic Chemicals and Public Protection.*

17. Ibid.

18. Toxic Substances Control Act, 15 U.S.C., sections 2601–2629 (1976).

19. Chemical substances covered under TSCA include most chemicals used commercially, with the exception of pesticides and chemicals regulated by the Food and Drug Administration (drugs, food additives, and cosmetics).

20. 15 U.S.C., section 2601(2)(b).

21. Luther J. Carter, "Toxic Substances: Five Year Struggle for Landmark Bill May Soon Be Over," *Science* vol. 194 (October 1976) pp. 40–42.

22. *Toxic Substances Control Act,* H. Rept. 1341 to accompany H.R. 1403, 94 Cong. 2 sess. (1976) pp. 14–15.

23. National Academy of Sciences, *Regulating Pesticides* (Washington, D.C., 1980).

24. Federal Insecticide, Fungicide, and Rodenticide Act, 7 U.S.C. (1947), as amended, 1972, 1975, 1978, 1980.

25. Ibid.

26. In 1988, Congress amended FIFRA again, establishing a mandatory nine-year schedule for reregistering pesticides.

27. This conflict was the subject of a recent National Academy of Sciences report, *Regulating Pesticides and Food: The Delaney Paradox* (Washington, D.C., 1987).

28. R. Shereff, "Bringing the New Hand into the Field," *Chemical Business*, February 1985, pp. 35–37.

29. ICF Incorporated, *Regulatory Impact Analysis for New Chemical Reporting Alternatives Under Section 5 of TSCA* (Washington, D.C., Office of Pesticides and Toxic Substances, Environmental Protection Agency, 1983).

30. A bioassay can be expensive and time-consuming. From initial planning to final data evaluation and interpretation, the study can take four years and cost about $800,000. Short-term, inexpensive tests have been developed that can be used to predict potential carcinogenicity based on the ability of chemicals to induce genetic changes in animal cells or bacteria. The more successful of these tests correlate very well with the results of bioassays and are quite inexpensive, ranging from about $1,000 to $20,000. (See National Academy of Sciences, *Toxicity Testing: Strategies to Determine Needs and Priorities* [Washington, D.C., 1984]; Office of Technology Assessment, *Assessment of Technologies for Determining Cancer Risks from the Environment* [Washington, D.C., 1981].) Such short-term tests are widely used as screening tools and to provide supportive evidence in carcinogenicity assessment. Additional research will enhance their value. However, at present a full bioassay must be conducted to obtain adequate data for a quantitative risk assessment.

31. Faced with the uncertainties, the EPA has developed generalized guidelines on performing each of the steps identified above, based on consensus among scientific experts. For example, EPA's guidelines presently call for using the linearized multistage model for low-dose extrapolation. The model basically produces a linear-dose response relationship at low doses and assumes that there is no exposure level below which the risk of getting cancer is zero (that is, there is no threshold effect). This model is generally thought to be conservative in that it predicts a higher probability of getting cancer than do most other extrapolation models.

32. Assessments of such effects are usually based upon determinations of no-observable-effect levels in animal studies, combined with safety factors or margins of safety for extrapolating to human no-effect levels. This methodology does not allow one to determine the relative degree of risk at different levels of exposure, but can be and has been used to set standards.

33. National Academy of Sciences, *Toxicity Testing.*

34. The National Toxicology Program (NTP) was established in the Department of Health and Human Services to further the development and validation of toxicological test methodologies. (See Toxic Substances Strategy Committee, *Toxic Chemicals and Public Protection.*) Federal regulatory and scientific agencies (including the EPA) participate in the management of the NTP and in the selection of chemicals for testing by the NTP. In contrast to TSCA test rules, NTP testing is funded by the federal government rather than industry.

35. The Office of Pesticides Programs has recently begun a reexamination of the inert

ingredients in pesticides, since in a number of instances they have proven to be toxic materials.

36. This is the annual contribution to a sinking fund at 10 percent interest for 10 years that would be needed to pay for the testing over a 42-year period. A figure of 100 chemicals per year was arbitrarily related to reflect limited capacity to test chemicals and review the data generated.

37. *Natural Resources Defense Council (NRDC) v. Costle,* 14 ERC 1958 (S.D.N.Y., 1980).

38. R. C. Evans, J. Bakst, and M. Dreyfus, "Analysis of TSCA Reauthorization Proposals," draft report, Office of Pesticides and Toxic Substances, Environmental Protection Agency, 1985.

39. Negotiated agreements between the EPA and industry have been a principal mechanism for getting testing done. However, because such agreements are not legally enforceable and are not specifically permitted by TSCA, a New York District Court ruled in 1984 that the EPA could not use voluntary agreements in lieu of rules in responding to ITC designations. (See *NRDC and AFL-CIO v. EPA,* 83 Civ. 8844, Slip op. at 18 [S.D.N.Y., 1984].) The EPA has subsequently promulgated regulations for implementing a negotiated testing process that would satisfy these legal objections.

40. Evans, Bakst, and Dreyfus, "Analysis of TSCA."

41. U.S. Congress, Senate, "A Bill to Amend the Toxic Substances Control Act," S. 3075, 98 Cong. 2 sess. (1984).

42. In 1986, the EPA also issued a minor regulation under TSCA extending Occupational Safety and Health Administration worker protection requirements for asbestos abatement to state and local government employees under OSHA jurisdiction. In July 1989 the EPA issued a final rule that would result in the elimination of most commercial uses of asbestos over a ten-year period.

43. "Facts and Figures," *Chemical & Engineering News.*

44. 48 Federal Register 52507 (1983).

45. ICF Incorporated, "Cost Analysis of Options for Controlling Formaldehyde Emissions from Wood Products," draft report, Office of Pesticides and Toxic Substances, Environmental Protection Agency, 1985.

46. 49 Federal Register 21870 (1984).

47. "Pesticide Pact Struck by Opposing Groups," *Science* vol. 230 (October 1985) pp. 48–49.

48. 44 Federal Register 2242 (1979); 48 Federal Register 21722 (1983).

49. In 1985 two exemptions became effective, one for polymers, the other for new chemicals with an annual production volume of less than 1,000 kg. These exemptions reduce the reporting requirements and review period for eligible chemicals; they do not entirely eliminate the need to report. (See 49 Federal Register 4606 [1985]; 50 Federal Register 16477 [1985].)

50. Office of Technology Assessment, *The Information Content of Premanufacture Notices* (Washington, D.C., Government Printing Office, 1983).

51. "Facts and Figures," *Chemical & Engineering News.*

52. The calculations presented here assume that the PMN substances would be commercially successful. In fact, about 40 percent of the PMNs submitted through the end of fiscal year 1982 have not been manufactured. Therefore, in evaluating the impact of testing, one should more realistically take into account the uncertainty regarding the future streams of sales.

53. "Toxic Substances Control Act," H. Rept. 1679 to accompany S. 3149, 94 Cong. 2 sess. (1976) p. 65.

54. P. Huber, "Exorcists vs. Gatekeepers in Risk Regulation," *Regulation* vol. 7 (November/December 1983) pp. 23–32.

55. B. F. Mannix, Letter to the editor, *Regulation* vol. 8 (March/April 1984) p. 3.

56. D. A. Leaf, "Acrylates: An Overview," draft report, Office of Pesticides and Toxic Substances, Environmental Protection Agency, 1985.

57. "Regulatory Development Branch Formed to Focus on Risks of Chemical Categories," *Chemical Regulation Reporter*, May 24, 1985, p. 211.

58. E. J. Heiden and A. Pattaway, *Impact of the Toxic Substances Control Act on Innovation in the Chemical Specialities Manufacturing Industry* (Washington, D.C., Chemical Specialties Manufacturers Association, 1982).

seven

Monitoring and Enforcement

Clifford S. Russell

How often does one hear it said, "In the United States, if a problem appears, all we ever seem to do is pass a new law"? In some cases this complaint may imply that the speaker has an alternative approach in mind. But in many situations what seems to be meant is that we are wont to pass laws and then lose interest in the results. What is missing in such situations is a commitment of resources to checking up on whether those covered by the law and regulations are doing (or not doing) what is required of (or forbidden to) them—this is *monitoring*; and to taking actions that force violators to mend their ways and that provide visible examples to encourage others in the regulated population to maintain desired behavior to avoid a similar fate—this is *enforcement.*

This chapter deals with monitoring and enforcement of the nation's environmental regulations. It argues that this area of public policy shows signs of the sort of half-hearted commitment just described. Efforts to monitor regulated behavior appear to have been inadequate to the task—a very difficult task in many instances—and typical enforcement practices appear to have been insufficiently rigorous. Together these inadequacies seem to have encouraged widespread violations of environmental regulations. Unfortunately, however, the very lack of monitoring means that the evidence on compliance or lack thereof is spotty at best and nonexistent at worst.

We begin by drawing distinctions among the different monitoring problems posed by the range of environmental regulation. Some of the very real difficulties in monitoring are then discussed, and to balance the critical comments, notice is taken of recent efforts at improvement. The chapter's final section offers several specific suggestions for policies—and even legislative actions—that should help make further progress easier. It will be useful to begin with some distinctions and elaborations on the simple definitions of monitoring and enforcement set out above.

DISTINCTIONS AND DEFINITIONS

The complexity and the technological bases of U.S. environmental regulation imply that several different kinds of monitoring and enforcement activities have claim to our attention. The air and water pollution legislation of the 1970s stressed the issuance of technology-based discharge or emissions standards. (As discussed by Freeman in chapter 4, this focus represented a reaction against the ineffectiveness of previous efforts to assert control on the basis of ambient quality standards or problems.) The practical result for monitoring and enforcement was concentration on determining that the required technology (pollution control equipment) was in place and capable of performing in such a way as to meet the terms written into discharge permits. This is known as *initial compliance* monitoring. Much of the discussion of compliance rates, successes, and failures over the past decade and a half has referred to initial compliance. Enforcement actions involving failures of initial compliance have been roughly of two sorts: penalties for failing to make any effort to purchase and install the technology; and denial of permission to operate if the equipment, once installed, fails to produce the desired results.

Clearly, however, the purpose of technology-based regulations ought to be to affect the actual discharges from the sources, and to do so on a continuing basis. Checking up on this involves monitoring for *continuing compliance.* Enforcement actions, in principle, follow from findings of failure to live up to permit terms, and may take many forms, from a slap on the wrist to a full-blown criminal prosecution, depending on how serious the violation is thought to be and how cooperative or aggressive the violator appears. Current environmental laws and regulations contemplate that a common form of enforcement action will entail the collection of an administrative or civil penalty designed to capture any cost saving reaped by the source from the violation, plus the extraction of something in the nature of a fine.[1]

The ultimate goal of the laws and regulations is to improve (or in some cases maintain) the day-to-day air and water quality experienced by citizens generally. This quality is the result of the interaction of the discharges of pollutants by sources with the natural systems that receive them—the atmosphere, water courses, land and underground aquifers. These systems dilute, transport, and even transform the pollutants they receive. The complexities of such changes in form, location, and concentration are so great that ambient conditions could not with any confidence be deduced from discharges alone (even if we knew discharges quite exactly—which we do not). Rather, ambient quality must be checked directly if we want to know what we are getting for our pollution control spending. This is *ambient quality* monitoring. The three types of monitoring and their relations to each other are shown schematically in figure 7–1.

This chapter concentrates on monitoring and enforcement of continuing compliance of individual sources. In passing, it will be necessary to cite evidence on initial compliance because this has been dominant in the Environmental Protection Agency's discussions of its activities. But very little will be said about ambient quality monitoring. Our focus should not be taken to imply either that ambient quality monitoring is easy or that it is being done as well as it might be. On the contrary, many observers and analysts argue that far too little effort is made in both air and water quality monitoring.[2] With respect to air quality, the number of monitoring stations and their placement implies a dearth of information on rural pollution levels; in the case of water pollution, roughly the opposite is true—there is a dearth of consistent records for the lower, often urbanized, stretches of rivers and their estuaries. Further, audits of ambient air monitoring stations have suggested that many of them serve their purpose poorly.[3]

Nor is ambient quality monitoring unimportant. One of its functions is to check actual conditions against the ambient standards that express society's judgments about what should be achieved (see chapters 3 and 4). Therefore, ambient monitoring is not simply information-gathering but ultimately is linked to violations and enforcement. The violations, however, are not in general attached to individual sources. Rather, a region or a water course may be found to be failing to meet its ambient quality targets. In the case of air quality, this makes the region a nonattainment area, and as such subject to a variety of special constraints and requirements.

Potentially at least as important, the ambient quality record is the sine qua non for judging the effectiveness of environmental regulation. While inferring effectiveness from such a record is far from trivial (see chapters 3 and 4), nothing can be done in the absence of such a record. Further,

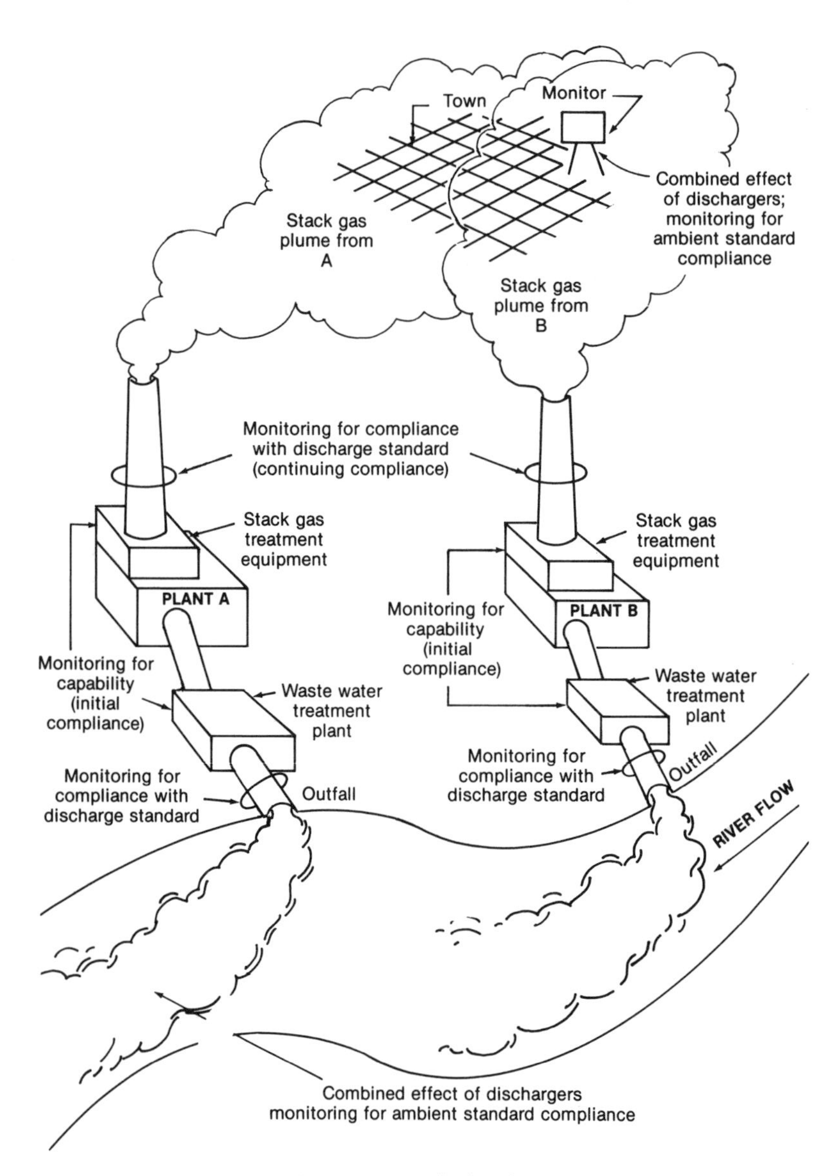

Figure 7–1. Schematic of monitoring distinctions

estimation of the benefits of environmental regulation—an activity important to better policy formulation in the future—almost invariably requires a record of actual environmental quality levels. This has to be matched with base-period data on health status or recreational activities, material damage levels, or agricultural yields in order to estimate statistically the relevant response functions.

The decision to concentrate here on continuing compliance monitoring reflects two judgments: first, that of the kinds of monitoring this has been relatively least discussed in policy analysis; and second, that success in encouraging continuing compliance by sources would produce ambient quality at or near the desired ambient standards.[4]

Within the narrower field of monitoring for continuing compliance, there are two other broad distinctions that are worth setting out. The first is that the quantities and characteristics of the pollutants being measured can differ from source to source (and even between two identical sources located in different jurisdictions). The differences include:

1. Types of pollutants measured. Nitrogen, for example, appears in different chemical states in both waste water and waste (stack) gases. In addition, in discharge permits the compounds being constrained may differ, and so may imply differences in monitoring difficulty.
2. Forms in which the materials enter the environment—as gases in a stack; compounds dissolved or suspended in waste water; hazardous materials in drums or in such spent consumer goods as old appliances; or mercury in batteries or PCBs in old electrical transformers. Again, different monitoring problems are implied. A smokestack, for example, is at a fixed point and visible to the public eye. The dumping of drums or transformers can go on anywhere at any time.
3. How the standard is defined. Alternatives include pollutant concentration (for example, milligrams per liter of a compound dissolved in waste water), mass per unit time, and mass per unit input to or output from the facility in question. Concentration alone is easiest to measure and least useful in protecting the environment. Mass per unit input or output involves the most elaborate measurements but may not provide the desired protection either, since permitted pollution may be allowed to increase with output.

The second broad distinction involves the time dimension. It may best be understood by example. On the one hand, a standard may specify that on *no* day are discharges to exceed 50 tons of SO_2; on the other, daily discharges *on average* over a month might be limited to 50 tons. The second specification does not constrain any particular day's discharge, for amounts of more than 50 tons discharged on some days may be balanced by amounts under that level on others (it does constrain any month's total discharges to be fifty times the number of days in the month, however). This distinction is important, among other reasons, because of its implications for the frequency of monitoring activity as related to inferences about source compliance rates. While this is a very complicated subject, an intuitive appreciation may be gained from the observation that if the responsible agency measures a source's discharge for one day per month, it will have direct evidence only about compliance or violation with respect to a single day's upper limit and only for that day. Inferences about other days and about compliance with monthly limits (daily averages over a month) involve strong assumptions about constancy of source behavior (the representativeness of the sample) *and* exogenous influences such as temperature, volume of production, and input quality.

A SKETCH OF THE CURRENT SYSTEM

For the U.S. environmental monitoring and enforcement system, one might generously describe the present period as one of transition. From the passage of the new clean air and water legislation in the early 1970s until the last few years, the EPA—and perforce the related state agencies—have been interested almost entirely in maintaining progress on the installation of technology required for meeting discharge permit terms. In the words of one EPA observer,

> The traditional approach to compliance monitoring and enforcement consists of periodic inspections by government employees, warning letters or notices for some or most violations, followed by additional warnings, or escalation to administrative orders or civil litigation and case referral. Some programs are assisted by information on source compliance achieved through source self-monitoring requirements, but under the traditional approaches, such self-monitoring is imposed more to ensure higher levels of compliance through source self-awareness than it is to manage an enforcement program.
>
> The traditional approach worked fairly well when compliance focused on initial installation of pollution control equipment at a limited number of large facilities. Detection was comparatively straightforward. A periodic inspection could do the job. Heavily publicized selective enforcement actions complemented what were then new pressures and norms of behavior for environmental protection for which there was a large degree of public consensus. Fixed hit lists of known violators proved to be effective in gaining compliance. The traditional approach has run into several problems . . . *In the older established programs such as air and water, the problems have shifted from initial compliance to continuing compliance.* . . . [emphasis in the original][5]

The concentration on initial compliance has had a pervasive influence on the system. Everything from technology design to the laws limiting the agencies' rights to enter a plant to monitor, and from the design of noncompliance penalties to the keeping of program statistics, has reflected it. In particular, monitoring equipment as approved by the EPA has been unwieldy and expensive to set up and operate. For its efficient use, skilled personnel have been required, as well as notice to the source and cooperation from the source. The goal, judging by the standards applied to newer technologies, has been to keep errors of measurement low.

Moreover, noncompliance penalties have been designed primarily to recapture the costs that polluters avoid by failing to meet requirements. Where capital goods purchases are in question, as they are in initial compliance, the largest part of these avoided costs are easily identified. Where only operating costs are involved, as they would be in cases of continuing compliance failures, measurement difficulties may be severe. It may be difficult, for example, even to identify the causes of inadequate performance by wastewater treatment plants, let alone to estimate the cost savings from allowing those causes to occur.

Further, those figures on the percentage of sources in compliance that have been regularly reported by the EPA—and that have shown very high levels of compliance—have historically been based on initial, not continuing compliance. Indeed, as will become clear below, it has been very difficult for the EPA or anyone else to know what levels of continuing compliance are in fact being maintained.[6]

This is not to say that no efforts have been made to ascertain and encourage continuing compliance. The major characteristics of the efforts that have been made are outlined below.

Heavy reliance on self-monitoring by sources. Almost all states require almost all large sources of air and water pollution to monitor their own discharges and report on the results. According to the results of a 1982 survey by Resources for the Future (RFF) of state officials responsible for compliance monitoring, only one of the responding states used no self-monitoring for existing sources.[7] In the case of air pollution, 4 of the 22 reporting agencies required all air pollution sources to self-monitor. Averaged over the responding states, 28 percent of air pollution *sources* were required to self-monitor. As regards water pollution, a self-monitoring requirement was almost universal. Of the 33 state agencies responding to the survey, two-thirds required all water pollution sources to do some self-monitoring. And the overall average percentage of *sources* required to self-monitor was 84.

Continuous self-monitoring of air pollution emissions is required of all facilities subject to New Source Performance Standards under the Clean Air Act. And even in the hazardous materials field, some self-monitoring is in force in the shape of a requirement under the Resource Conservation and Recovery Act (RCRA) that all generators of hazardous materials keep track of their wastes after the wastes leave the plant (by using the complicated shipping manifest system) and report if the wastes fail to reach an approved final disposal facility. Self-monitoring of groundwater around land disposal sites is also required under RCRA.[8] (See chapters 4 and 5 for further discussions of self-monitoring.)

Infrequent auditing of the self-reporting sources.[9] Results from the RFF survey show that while large sources are audited more frequently than small ones, on average neither is visited very often (see table 7–1), especially given the fact that the majority of the standards in question involved hourly and daily emission limits.[10] Thus outside measurement of the discharges of large air pollution sources occurs on average only once every eight and one-half months. For water pollution the frequency is greater, but still notably low—once every five months.

Table 7–1. Frequency of Audit Visits to Self-Monitored Sources

	Type of source			
	Air pollution		Water pollution	
Type of visit	Large sources	Small sources	Large sources	Small sources
Without measurement				
Responses (number)	26	15	41	35
Mean (times/yr)	1.70	0.88	3.10	1.44
Standard deviation	1.19	0.64	3.21	1.71
Including measurement				
Responses (number)	18	10	34	26
Mean (times/yr)	1.43	0.89	2.41	1.39
Standard deviation	1.26	0.81	2.91	1.72

Source: Clifford S. Russell, "Pollution Monitoring Survey: Summary Report," (Washington, D.C., Resources for the Future, 1983) p. 16. See note 7 to this chapter.

Audit visits lack the characteristics of a rigorous enforcement effort designed to catch ongoing violations. Most important, those visits that do occur are often announced in advance (see table 7–2).[11] To some extent this is a response to the uncertain state of the law regarding entry for monitoring purposes, in combination with the technological limitations of the approved monitoring methods. Whether the Environmental Protection Agency has the right to make unannounced, warrantless visits for monitoring purposes is an unsettled question.[12] Announcement helps to avoid quarrels and possible litigation about the right of entry.

A related legal question was recently decided by the U.S. Supreme Court, although by a 5 to 4 majority, and with important distinctions drawn that may well encourage future litigation. In this opinion it was found that the remote monitoring instrumentation used by the EPA did not give rise to an unreasonable search. This was a long-running case involving aerial photography of a Dow Chemical plant to check for sources of air pollution, and the District and Appeals courts had disagreed. The Supreme Court, in May 1986, agreed with the Sixth Circuit Court of

Table 7–2. Auditing Self-Monitoring Sources: Conduct and Content of Visits

	All sources	Air pollution sources	Water pollution sources
Conduct of audits			
Responses	$N = 68$	$N = 26$	$N = 42$
Always announced	0.16	0.19	0.14
Never announced	0.19	0.19	0.19
Sometimes announced	0.65	0.62	0.67
Sum of frequencies	1.00	1.00	1.00
Content of audits[a]			
Responses	$N = 68$	$N = 25$	$N = 43$
Inspect records	0.97	0.96	0.98
Inspect equipment	0.93	1.00	0.88
Measure discharges[b]	0.90	0.80	0.95
Other[c]	0.12	0.04	0.16

Source: Russell, "Pollution Monitoring Survey: Summary Report."

[a] The sum of the frequencies for items describing audit contents need not sum to one.

[b] The difference between air and water proportions is significant at 5 percent or better.

[c] Other activities undertaken as part of audits consisted almost entirely of one or another version of a laboratory inspection and a check on analytical methods. Frequencies do not sum to one because in principle every audit could include every content item.

Appeals and upheld the EPA's right to use aerial surveillance for enforcement. This should strengthen the general case for remote monitoring, with implications for the entry question itself. However, the majority opinion drew a distinction between use of a standard commercial camera and "highly sophisticated . . . equipment."[13]

A certain amount of ad hoc invention in the definition of violations. Even though emissions standards are clearly stated under outstanding permits, and even though those standards are based on official documents that take into account discharge variations resulting from changing circumstances beyond the control of the source, another layer of allowance has been tacked on to separate "significant" from "insignificant" violations. These allowances have both a quantitative and a time dimension. Discharges must exceed the actual standard by some amount for some period of time before a significant violation will be said to have taken place.[14]

Infrequent and reluctant use of self-monitoring records as the basis for notices of violation, even though the records show significant violations to be occurring. Thus the General Accounting Office (GAO) observed that, on the basis of a study of water pollution self-monitoring records in six states, formal enforcement action was in some cases not taken for years after noncompliance began.[15]

Indeed, some major environmental groups, feeling that enforcement had become too lax, began a campaign in 1983 to use self-monitoring records—which are public information—as the bases for citizens' lawsuits against the violating polluters.[16] These suits have been successful in at least two ways. First, the courts have ruled against the polluters on most of the technical legal challenges the latter have raised; thus, as part of the overall enforcement armory of society, the device has been found valid under the existing laws.[17] Second, and no doubt more important, the suits have helped embarrass the EPA into rethinking its own enforcement stance, as will be discussed below.

When violations are found and enforcement actions taken, the penalties assessed appear to be so small as to be insignificant in a corporation's (or city's) income statement. Many states claim to pursue a so-called voluntary compliance policy, by which they mean that no penalties are ordinarily levied for violations initially. Rather, if penalties are used, it is to punish sources that refuse to correct violations or otherwise prove notably uncooperative. Even when penalties are assessed, however, they seem on average to be very small. Some evidence of this is presented in table 7–3, which shows both the average penalty per assessment and the average penalty per notice of violation issued. The latter figure can be thought of as the expected penalty per discovered violation. This expected value of the penalty for a violation is closer to the incentive that can be presumed to drive the source's compliance decision than is the average penalty per assessment.[18] The EPA's official civil penalty policy, as mentioned above, is to assess a violator a two-part penalty.[19] The first part is an attempt to remove the profit from violations by recapturing the costs saved by the source. The second part, or gravity component, is intended as punishment pure and simple. This policy has not been in effect long enough to have produced data on its impact.

Before examining the evidence on continuing compliance, such as it is, from the system just described, it is worth adding that at least one other major industrial nation appears to have a similar system. Papers and books about the United Kingdom's activities in pollution control monitoring and enforcement paint a roughly similar picture.[20] Thus in the United Kingdom self-monitoring is important; agency personnel visit infrequently and measure discharges independently even less frequently; and much stress is laid on a version of voluntary compliance, with penalties reserved for recalcitrant violators. The reasons suggested for the evolution of the U.K. system in this direction are about what would be expected: low agency budgets; legal difficulties of bringing a serious prosecution for a "mere" violation of a discharge standard; and the unquantifiable, personal considerations that push toward amiability rather than confrontation when inspectors and plant management know they must live together over the long run.

EVIDENCE ON CONTINUING COMPLIANCE

Since one of the characteristics of the existing system is that very little monitoring is done by the Environmental Protection Agency or its state counterparts, it is inevitable that very few data on actual continuing compliance exist. Some notion of the current situation can be gained from special studies, most frequently those requested by congressional committees and undertaken by the U.S. General Accounting Office.

Air Pollution

In the late 1970s the EPA and the White House Council on Environmental Quality (CEQ) sponsored jointly or singly nine studies examining continuing compliance with air pollution emissions limits by industrial sources.[21]

Table 7-3. State Enforcement Activity (annual averages, 1978–1983)

				Penalties ($)		
State	NOVs issued (1)	Civil actions brought (2)	Number of penalties assessed (3)	Total penalties assessed (4)	Average size of penalties assessed (5) = (4)/(3)	Average penalty per NOV (6) = (4)/(1)
Colorado	124	3.6	0.5	60	120	0.5
Connecticut	800	2.3	21.5	7,800	363[a]	9.8
Indiana	59[b]	NA	21.0	85,000	4,050[c]	1,440
Kentucky	194	5.2	5.2	13,100	2,520[d]	68
Massachusetts	NA	0	0	0	0	0
Minnesota	41	NA	10	109,000	10,900	2,660
Nebraska	59[e]	NA	0.2	400	200	0.07
Nevada	31.5	0.3	2.3	105	45	3.3
New Jersey	1,167	NA	350	500,000	1,430	428
Oregon	197	NA	30.7	21,600	705	110
Pennsylvania	NA	NA	176[f]	260,000	1,480	NA
Rhode Island	5	7.2	0	0	0	0
South Carolina	68[g]	5.5	2.2	53,400	24,250[h]	785
South Dakota	17	1.2	0.3	300	1,000	20
Tennessee	193[i]	8.4	0	0	0	0
Virginia	161	7.8	3	600	200	3.8
Wisconsin	80.5[j]	13.5	7.7	61,200	7,951	760

Note: NOV = Notice of Violation; NA = not available.

Source: Clifford S. Russell, Winston Harrington, and William J. Vaughan, *Enforcing Pollution Control Laws* (Washington, D.C., Resources for the Future, 1986) table 2-7.

[a] Refers to amounts assessed; over the same period actual collections were 62 percent of assessments.

[b] Includes both NOVs and Compliance Orders.

[c] Excludes one penalty of $415,000, which was cancelled when company bought equivalent amount of air pollution equipment.

[d] Excludes performance bonds (two required for total of $45,000).

[e] Excludes Lincoln and Omaha.

[f] Only data from the last quarter of 1983 were readily available. This figure is an extrapolation to an annual average.

[g] No NOVs were issued before April 18, 1983. From July 1, 1983, to March 31, 1984, 51 NOVs were issued.

[h] Excludes one fine of $1,700 per month until compliance was restored and one fine of $250,000 dropped in consideration of a donation of a like sum to a technical college.

[i] Does not include NOVs from continuous monitoring data, which were numerous.

[j] 1980–1983 only.

Two tasks were undertaken. First, state and local agencies were studied and their procedures for pursuing continuing compliance monitoring and enforcement were analyzed. Second, data were gathered on raw pollutant loads, types of control equipment, emissions, and causes of excess emission "incidents." The data from the second task give some (albeit very modest) insight into compliance rates among a small sample of 119 sources from the 23,000 large sources estimated to exist in the United States. Within these limitations, the following data are relevant: percentage of sources in violation—65; percentage of time the sources were in violation—11; excess emissions as a percentage of standards—10.[22]

Before considering what can be learned from the summary document for these studies, it is worth noting why the effort is less valuable than it otherwise might be. To begin with, discussion in the summary document does not make clear what standards form the basis for the "allowable" and "excess" emissions calculations. It is implied that the basis is permitted daily emission limits, or perhaps permitted hourly limits. But the calculation of a "credit" for discharge levels below the total annual permitted amount suggests that it is long-term average performance that is really of interest. While the data are said to be for emissions of various pollutants, no breakdown of incidents or standards by pollutant is offered.

In addition, in the summary document the sources of data are only vaguely described as inspections and records. It does not seem that independent source measurement was involved, suggesting that the results at least do not overstate the size or frequency of violations. Moreover, the causes of violation allowed for in the studies do not include deliberate evasion. Equipment misdesign or malfunction, excusable process accidents (such as a burst pipe or burned-out transformer, beyond operator control) are the only possibilities allowed for. The result is a catalog of "blameless" events.[23]

An even less satisfactory source of data on compliance with air pollution emission standards is provided by a 1979 report from the General Accounting Office.[24] For this report, the GAO staff reviewed documents in the EPA regional offices and in state and local air pollution control agencies. No independent measurements of emissions were undertaken. The principal problem with the report, however, is that it is very unclear on the distinction between continuing and initial compliance. To the extent one can infer what is being discussed from the imprecise language, it appears to deal with initial compliance. A major message of the report does have some relevance for the discussion of continuing compliance, however, because it confirms the characterization of the system given above. The GAO reports that actual emissions monitoring, and even on-site inspection without such monitoring, are very seldom used to determine compliance status. (Only 3 percent of the sources reported to be in full compliance were so characterized on the basis of source tests; only 22 percent had been inspected without a source test.) Reinspection of sources discovered in violation was also found to be spotty. And 50 percent of the sources reported to be on clean-up schedules were found to be in violation of those schedules. A more recent report by the GAO found that in 1984, while some inspection was performed on 95 percent of stationary sources, 40 percent of the inspections performed were inadequate on grounds of either content or infrequency of repetition.[25]

Water Pollution

The General Accounting Office has also prepared two reports on continuing compliance with water pollution control permit terms.[26] These are somewhat more useful than either of the air pollution reports. Though both reports on water pollution control share the understandable weakness of being based entirely on reviews of self-monitoring records, they do at least focus clearly on month-to-month behavior and do supply clear definitions of compliance status. Indeed, since both reports share the same definition of significant noncompliance, it is even possible to make a simple comparison across the space of time separating the reports.[27] Both reports are based on data from major sources, defined as those discharging more than one million gallons of treated water per day.

The total number of incidents of violation found by GAO investigators is shown in the first column of table 7–4. The second, third, and fourth columns show how the total breaks down by length of period of violation: short (one to three months), medium (four to six months), and long (more than six months). It appears that short periods of violation were just as common in 1980/82 as in 1978/79. But long periods of violation, greater than six months, were much less common in the later period. This is an encouraging indicator. On the other hand, when account is taken both of the length and severity of violations, as the GAO did when devising a measure of significant noncompliance, the picture is darker (see table 7–5). At best it would seem that significant noncompliance remained constant between the two periods among large municipal sources. The data do not permit an intertemporal comparison for industrial sources, but the 1980/82 data do indicate that such sources are reporting less significant noncompliance than are municipal treatment plants.[28] This is especially interesting because most of the attention given to pollution problems and violations in the media focuses on industrial sources.

Table 7–4. Extent of Violations Found in Self-Reported Data

Report	Total incidents of violation	Months in violation of at least one limit[a]		
		1–3	4–6	more than 6
1978/79				
Municipal				
N = 242	211	53	39	119
(%)	(87)	(22)	(16)	(49)
1980/82				
Municipal				
N = 274	139	44	29	66
(%)	(51)	(16)	(11)	(24)
Industrial				
N = 257	135	63	24	48
(%)	(53)	(25)	(9)	(19)

Sources: General Accounting Office, *Costly Waste Water Treatment Plants Fail to Perform as Expected*, CED-81-9 (Washington, D.C., 1980); General Accounting Office, *Waste Water Discharges Are Not Complying with EPA Pollution Control Permits*, RCED-84-53 (Washington, D.C., 1983).

[a] For 1980/82, where separate figures were given for concentration and quantity violations, this table gives results for quantities.

The GAO reports on water pollution control include some information on the causes of observed violations, as determined by discussions with knowledgeable state and EPA officials. In the 1978/79 study, five major causes were identified: equipment deficiencies, intake volume overloads, operation and maintenance deficiencies, design deficiencies, and intake pollutant overloads or toxicity. The 1980/82 study found the first three of these were the most prevalent causes of violations in that period. Once again, it is interesting that no explicit account was taken of the possibility of deliberate violation.

It is also worth noting that in the 1980/82 study the General Accounting Office criticized state and EPA enforcement practices for allowing violations to continue uncorrected for long periods of time. It appears that any kind of formal enforcement action may be delayed for months or perhaps years, even after a significant violation is revealed by self-monitoring reports. This may be because taking the enforcement action far enough to assess a civil penalty adds substantially to the required time,

Table 7–5. Significant Violations of Water Pollution Discharge Permits

Report	Plants
1978/79	
Municipal	
N = 242	66
(%)	(27)
1980/82	
Municipal	
N = 274	88
(%)	(32)
Industrial	
N = 257	42
(%)	(16)

Note: Significant violations are defined by size of violation and number of periods the violation continued.

Sources: General Accounting Office, *Costly Waster Water Treatment Plants;* General Accounting Office, *Waste Water Dischargers Are Not Complying.*

since the Justice Department has to become involved. Indeed, the GAO concludes that "EPA's current enforcement philosophy for water pollution control still centers around voluntary compliance and the nonconfrontational approach established in 1982."[29]

Air Pollution by Automobiles

Congress has written specific discharge standards for emissions of automobile pollutants (hydrocarbons, carbon monoxide, and nitrogen dioxide) into the Clean Air Act (see chapter 3). The monitoring and enforcement effort for these standards originally was aimed only at the end of the assembly line, where brand new cars were to be tested—an effort analogous to determining initial compliance for stationary sources. As matters have subsequently developed, continuing compliance is now to be monitored through vehicle emissions tests, but only in 30 of 50 states and even there usually only at fixed intervals (an analog to announced monitoring visits).[30]

Two General Accounting Office reports—one issued in 1979 and the other in 1985—on the so-called inspection and maintenance (or I/M) program for cars provide other analogs to experience with stationary source control programs. Under the initial program of prototype testing and certification, very little effort was put into continuing compliance monitoring. The result, as determined by the EPA in the period 1972 to 1976, was that of 2,000 automobiles in use and sampled at that time, about 80 percent failed to meet the exhaust standards for their model year.[31] The readiness with which the failure to meet standards was ascribed by the EPA to conscious cheating by car owners stands in interesting contrast with the stationary source analyses reported above. In particular, it was estimated by the EPA that almost half the vehicle failures were due either to tampering with the pollution control equipment or to improper driving or fuel use. "Maladjustment," the single largest cause of failure, corresponds to operation and maintenance deficiencies in the stationary case. Apparently, in the government's view, individuals may cheat on pollution control for their vehicles but corporations or municipalities will not do so at the plants they operate.

In its 1979 report the GAO recommends expanded and improved monitoring for continuing compliance—at the least, annual vehicle inspections and necessary reinspection to check that needed work has been done. By the time the 1985 GAO report on this subject was issued, such programs were required in 44 metropolitan areas in 30 states and were actually in place in 25 or more.[32] The 1985 report criticizes the in-place programs for ineffective monitoring and enforcement, though the criticisms are not so harsh as those given the water program, and the EPA's efforts to address the problems are recognized. The major criticism in the 1985 GAO report is that enforcement has often been inadequate to spur owners to have their cars inspected. It was entirely too good a gamble from an owner's point of view to ignore the inspection requirement. In addition, some fraud in the issuing of "pass" stickers was observed in undercover visits to inspection stations.

A second criticism might or might not be valid. The problem identified by the GAO was that "too many" cars were passing inspection in some states. The benchmark for this conclusion was the EPA's judgment about how many cars would have to fail and be repaired in order for the program to have the desired effects on emissions and ambient air quality. These projections, however, could only have been made on the basis of the violation rates observed in the pre–I/M small sample checks done in the early 1970s. Whether those rates of violation would continue into an era having an active I/M program is at least doubtful. It may be, therefore, that the program has had the desired effect on continuing compliance, so that the pass rates would be "correct."

In a subsequent (1986) report on the vehicle inspection and maintenance program, the General Accounting Office commented on the EPA's responses to the 1985 report and brought some of the numbers up to date.[33] For example, by the time of the 1986 report, 42 of 44 areas required as of 1982 to produce I/M programs had either started the programs or had produced plans to start such programs by 1987. Until more is known about attainment of ground-level ozone standards, it will not be known how many other areas may have to put I/M programs in place under current regulations. But the EPA has hinted at the need for a nationwide program because of expected deterioration in the performance of the emissions control systems on 1981 and later cars. Some of the same enforcement problems noted in the 1985 report continued to exist in 1986, by the EPA's own admission. The reason appears to be the EPA's unwillingness to push hard for programs that are so unpopular with states and individuals.

Hazardous Waste Disposal

If the problems of checking up on the performance of stationary sources of conventional pollutants have proved difficult for the responsible agencies, it is reasonable to expect even greater difficulties in monitoring and enforcing the laws and regulations designed to produce correct disposal of hazardous materials. Hazardous materials are often produced in small quantities in separable streams within production processes. Thus hazardous materials are often transportable, so that the point of final disposal in the environment is not fixed as it is for waste water and stack gases. The aim of existing legislation is to make sure that the point of final disposal is a carefully designed and regulated facility. The disposal options that will be acceptable constitute a rapidly shrinking universe. Under the RCRA amendments of 1984, all land disposal of untreated wastes will be banned by the beginning of the 1990s, while deep-well injection will only be acceptable if the operator can demonstrate that the wastes will not migrate from the injection zone for 10,000 years or that they will no longer be hazardous by the time they do migrate. The intention is clearly to make it difficult or impossible to contemplate any waste management alternatives other than source reduction, recycling, or high-temperature incineration.[34]

The monitoring system chosen bears some resemblance to that adopted for stationary sources. In particular, self-monitoring is imposed for waste transportation in the form of a system of manifests; and for groundwater contamination by requirements for numbers of test wells, well placement, and water test frequencies. Any generator of a waste who is incapable of disposing of that waste in an acceptable manner on site must fill out a manifest giving details of the amount and characteristics of the waste. This paper accompanies the waste when it is shipped off site, and a copy is supposed to be returned to the generator signed by someone at the authorized disposal facility where the waste ends up. The self-monitoring feature is most clearly seen in the requirement that the generator report to the responsible agency any anomalies, such as nonreceipt of a manifest for a particular shipment.

On its face, such a system would seem open to abuse if sources or shippers have deliberate evasion in mind. For example, a generator of hazardous wastes could dispose of the wastes illegally, thus avoiding the manifest system altogether. Further, auditing the self-monitoring operation would be a considerable job, given the large number of individual sources involved. (It has been estimated that 175,000 small sources of waste were added to the regulated population by the 1984 amendments to RCRA; as a group, however, they add less than one-half of one percent to the quantity of hazardous wastes generated each year.) The requirements generally are too new to have produced a long record of monitoring and enforcement success or failure. And of course the intrinsic logical difficulty of deciding how well a self-monitoring system is working applies here, too. But once again a General Accounting Office study provides some hints.[35]

One finding of this study (issued in 1985) has major implications for the long-term future. It is that of 36 cases of illegal disposal discovered and pursued as enforcement matters in the four states examined by the GAO study team, none were initiated on the basis of the manifest system or related inspections. Thirty-four were the product of information supplied by employees of the firms involved or by concerned citizens. Two were accidentally discovered during other investigations. Thus it should be no surprise that the GAO report is largely a catalog of speculations about the possible extent of illegal disposal and about the expected weaknesses of the monitoring system.

As far as enforcement actions go, in the 36 cases the GAO examined, 28 cases had been completed; in each of those 28, the state had obtained a favorable court opinion. In 10 cases prison sentences were imposed, though in 2 cases that part of the sentence was suspended. In 17 cases, fines were levied, supplemented in 7 cases by damage assessments. In 2 other cases, damages alone were assessed.[36] The fines averaged 30 percent of the maximum allowable penalties in the cases in which they were imposed. State officials told the GAO study team that the penalty record would deter sources from future violations, but one must be skeptical, given the weaknesses in monitoring and hence what must be a low probability of detection.

As for the enforcement of requirements on the handling of hazardous waste at its place of generation, or where it might be subsequently stored or disposed of, the evidence once again, though sparse, suggests problems. A 1987 GAO report characterizes the inspections done on these facilities pursuant to the 1984 RCRA amendments as neither thorough nor complete.[37] And the number of inspections (11,785) performed in fiscal year 1986 suggests that fewer than 10 percent of the facilities that should have been inspected (most annually) were in fact reached at all. Evidence for the charge that inspections were of less than desirable quality came from two tests. In one, 26 actual field inspections were evaluated by experts from the EPA office in charge; only one of these was judged thorough and complete. Overall, for the other inspections, more than 50 percent of the class I violations (those representing an actual or potential release of hazardous waste to the environment) were missed by the inspectors from state agencies, contractors, or the EPA regional offices. The second test consisted of a comparison of sequences of inspection reports for 42 waste handlers. The aim was to see whether earlier inspections missed items—such as the absence of a waste analysis plan—that later inspections turned up and that could not have arisen in the interval between inspections. In ten of the cases reviewed, violations were missed—a total of 95 class I violations. Generally, the report's conclusion is that the very newness of the program, and thus of the inspectors, is a large part of the problem. The major recommendation is for better training of inspectors.

One should add that in at least one respect these inspections do not look so bad. Of the 122 class I violations missed in 25 inspections, only 4 involved the major physical facilities such as tanks, waste piles, and surface impoundments most closely related to actual waste containment or disposal. Twenty others involved the "use and management of containers," while the rest were more in the nature of administrative failures, such as a lack of general written standards or plans.[38]

PRESENT EFFORTS TO IMPROVE MONITORING AND ENFORCEMENT

One should not conclude from this record that the Environmental Protection Agency is oblivious to or cynical about the problems of monitoring and enforcing in its areas of responsibility. To be sure, there was a widely shared perception of slackening zeal during the early years of the Reagan administration.[39] This perception was attacked directly by William Ruckelshaus as part of his campaign to change the agency's, and not incidentally the administration's, environmental image in the period before the 1984 election. For example, in September 1983 a task force on monitoring was set up by the EPA's deputy administrator.[40] While the charge for this group focused on ambient monitoring, enforcement needs were not ignored.

More to the point, Ruckelshaus made explicit his intention to take a tough line against violators. In speeches to the EPA enforcement staff in the summer of 1983, he referred to them as "pussycats" and indicated he wanted them instead to be seen as the "gorilla in the closet"—the bogeyman that state officials could use as a threat in their dealings with recalcitrant polluters.[41] That this kind of personalized pressure had some effect seems clear from the increase in enforcement actions recorded in fiscal year 1984 and later. In table 7–6, the EPA's civil referrals to the

Table 7–6. EPA Civil Referrals to the Department of Justice, Fiscal Years 1972 Through 1988

Fiscal year	Air	Water	Hazardous waste	Toxics, pesticides	Total
1972	0	1	0	0	1
1973	4	0	0	0	4
1974	3	0	0	0	3
1975	5	20	0	0	25
1976	15	67	0	0	82
1977	50	93	0	0	143
1978	123	137	2	0	262
1979	149	81	9	3	242
1980	100	56	53	1	210
1981	66	37	14	1	118
1982	36	45	29	2	112
1983	69	56	33	7	165
1984	82	95	60	14	251
1985	116	93	48	19	276
1986	115	119	84	24	342
1987	122	92	77	13	304
1988	86	123	143	20	372

Source: Based on *Mealey's Litigation Reports: Superfund* vol. 1, no. 18 (December 28, 1988) p. c-1.

Department of Justice over the period fiscal year 1972 through fiscal year 1988 are summarized by problem category. The dip in activity during the early Reagan years is obvious, as is the higher level of activity in the years 1984–1988. These higher levels were only slightly higher than those achieved during the notably environmentalist Carter administration, however. A similar message is conveyed by table 7–7, which shows administrative actions by the EPA under the six major environmental laws. The record of administrative actions runs from 1972 through 1988.

A second line of effort within the EPA has been to improve enforcement by defining a uniform and improved civil penalty policy. This effort culminated in a document distributed within the agency in February 1984, which set out a policy designed to be flexible and to promote the goals of deterrence, fair and equitable treatment of the regulated community, and swift resolution of environmental problems.[42] The general structure of the penalties contemplated under this policy has already been briefly described: the benefit component is calculated to remove the cost advantage enjoyed by any noncomplying source; and the gravity component is intended as a penalty tied to the seriousness of the damage resulting from noncompliance. It should be noted that while calculating the savings to polluters from noncompliance is difficult, the problem of attaching dollar penalties to the gravity of a violation is even harder than assessing the monetized damages of the incident. This is because a successful violation can be presumed to have some incentive effects on other polluters, making them slightly more likely to violate the rules themselves. The EPA document goes on at some length about how to determine the *relative* gravity of two or more incidents, but offers no guidance whatever on a benchmark; that is, the relative scale is not anchored to any dollar number. It will be interesting to see what is made of this guideline in practice.

A third line of effort in enforcement—undoubtedly the most dramatic if not the most important—is a new stress on the criminal prosecution of polluters. Since available penalties for conviction include time in prison (which might be characterized as the ultimate gravity component) and since the effort has been accompanied by a strong public relations exercise, it appears that the EPA wants to plant fear in the hearts of executives considering whether or not to comply with environmental regulations.[43] The spectre of the penalty is apparently seen as important enough to justify the added difficulty of prosecution: obtaining an indictment from a grand jury, establishing intent, and proving guilt beyond a reasonable doubt.

While the public stress on criminal prosecutions is a recent phenomenon, the power to bring such actions is as old as the major environmental laws.

Table 7–7. EPA Administrative Actions Initiated (by Act), Fiscal Years 1972 Through 1988

	Clean Air Act (1970)	Clean Water & Safe Drinking Water acts (1972/1974)	Resource Conservation & Recovery Act (1976)	Superfund (CERCLA) (1980)	FIFRA[a] (1947)	Toxic Substances Control Act (1976)	Totals
1972	0	0	0	0	860	0	860
1973	0	0	0	0	1,274	0	1,274
1974	0	0	0	0	1,387	0	1,387
1975	0	738	0	0	1,641	0	2,352
1976	210	915	0	0	2,488	0	3,613
1977	297	1,128	0	0	1,219	0	2,644
1978	129	730	0	0	762	1	1,622
1979	404	506	0	0	253	22	1,185
1980	86	569	0	0	176	70	901
1981	112	562	159	0	154	120	1,107
1982	21	329	237	0	176	101	864
1983	41	781	436	0	296	294	1,848
1984	141	1,644	554	137	272	376	3,124
1985	122	1,031	327	160	236	733	2,609
1986	143	990	235	139	338	781	2,626
1987	191	1,214	243	135	360	1,051	3,194
1988	224	1,345	309	224	376	607	3,085

Source: Based on *Mealey's Litigation Reports: Superfund* vol. 1, no. 18 (December 28, 1988) p. C-5.
[a] FIFRA = Federal Insecticide, Fungicide, and Rodenticide Act.

Actual use of the power built slowly during the early 1980s.[44] An article in the *Wall Street Journal* in January 1985 brought the EPA's tiny criminal investigation unit to national attention, and seems to have been the opening shot in the public relations effort.[45] At the national level, the campaign has included attempting to beef up the image of the unit within the Justice Department that does the actual prosecutions on evidence obtained by EPA criminal investigation.[46] But reports of actions by state and local environmental agencies on their use of criminal enforcement powers are also notably frequent in the press.[47] As of the fall of 1988, 460 individuals or corporations had been indicted for criminal violations of environmental laws, resulting in more than 300 convictions and the imposition of over 200 years of jail time, of which more than a quarter were to be served in federal prisons.[48] Table 7–8 summarizes the EPA's criminal enforcement efforts over the period fiscal year 1982 through fiscal year 1988, including referrals to the Department of Justice, prosecutions, numbers of defendants charged and convicted, the conviction rate, and information on the sentences handed out and served.

Even if the Environmental Protection Agency and related state agencies are given high marks for improving their images, restructuring and codifying enforcement penalties, and increasing deterrence by using criminal actions and the threat of jail terms, it is still fair to ask whether other strategies are also available. It is especially important to ask this question in the light of the unsettling evidence on continuing compliance; that is, even if discovered violations are uniformly and sharply dealt with, actual deterrence may be negligible since it depends as well on the probability of detection. The evidence on which to base a judgment about that probability suggests it must be quite small. Therefore, not only enforcement but also the prior business of monitoring deserve attention. Ideas for improvements in both areas are taken up in the next section.

SUGGESTIONS FOR FURTHER IMPROVEMENTS

There are two major paths to an improved system of monitoring and enforcement, both of which deserve serious attention from the EPA, the states, and even the U.S. Congress. One path concentrates on improving agency monitoring (or auditing) by making it less likely that a source will be able to tinker with discharges before measurement by the agency can take place. The second attempts to forge a tighter link between monitoring and enforcement and by so doing allow more efficient use of limited resources.

Table 7–8. EPA Criminal Enforcement Activities, Fiscal Years 1982 Through 1988

	1982	1983	1984	1985	1986	1987	1988
Referrals to the Department of Justice	20	26	31	40	41	41	59
Cases prosecuted	7	12	14	15	26	27	24
Defendants charged	14	34	36	40	98	66	97
Defendants convicted	11	28	26	40	66	58	50
Conviction rate (%)	78.0	82.4	72.2	100.0	67.3	87.9	51.5
Months sentenced			6	78	279	456	278
per convicted defendant			0.23	2.0	4.2	7.9	5.6
Months served			6	44	203	100	185
per convicted defendant			0.23	1.1	3.1	1.7	3.7
Months probation		534	552	882	828	1,410	1,284
per convicted defendant		19.1	21.2	22.0	12.5	24.3	25.7

Source: Based on *Mealey's Litigation Reports: Superfund* vol. 1, no. 18 (December 28, 1988) p. c-3.

The first path has, itself, two major components: technology and law. Current monitoring technology was developed to determine initial compliance status. Initial compliance involves a well-defined situation in which *capability* is being tested. The source is required to run its production equipment at particular levels and to demonstrate that its pollution control equipment can produce acceptable discharges—discharges consistent with its permit. Advance notice does not give the source any advantage because the test does not involve actual operations. Indeed, notice may be required as a matter of fairness. And there is certainly no need for speed in setting up the measurement equipment. In this setting, complex and awkward measurement technology could be tolerated in the interests of keeping measurement errors low.

The same is not true when continuing compliance is to be monitored. In that context, ease with which equipment can be transported, brevity of set-up time, and economy of operation are more important than small standard errors. For if surprise in inspection can be achieved, multiple samples can be taken to reduce the overall error of the average measured discharge. The ultimate weapon in this regard is remote reading technology, with which no entry onto private premises need be made. Thus the first suggestion for improvement is:

- Concentrate monitoring research money on the development of simple and inexpensive instruments, and especially remote reading methods (such as those based on laser technology) that are capable of measuring stack emissions from outside the factory or utility gates.

The legal component for more effective auditing arises out of the need of state and federal pollution control agencies to have the right to enter private premises unannounced and warrantless for the purpose of measuring pollution discharges. This right is now in place for inspectors of nuclear power plants and mine safety. It seems reasonable that society's commitment to preventing conventional air or water pollution or hazardous waste contamination ought to be as firm as its commitment to preventing accidents at nuclear plants or mines. However, to achieve such an explicit commitment may require legislative action, for the state of the law as determined by the courts is currently unclear. (This same kind of legislative action is needed in the case of remote monitoring equipment as well, even though the Dow case mentioned earlier was decided in the EPA's favor.) Since the constitutionality of surprise inspections in analogous situations does not appear to be in question, it should be possible to achieve the necessary effect by amending existing laws to remove any doubts about the intent of Congress. Thus the second recommendation is:

- Amend as necessary all major environmental statutes to make it clear that Congress contemplates and approves of unannounced and warrantless visits by state or federal agency personnel for the purpose of making independent measurements of discharges. This approval should be explicitly extended to the operation of remote-sensing monitoring instrumentation. There should be no anticipation of privacy in the generation and discharge of pollutants.

The second major path toward improvement is to tighten the linkage between monitoring and enforcement. This could be accomplished straightforwardly by making the future frequency of monitoring depend on the past record of compliance for all significant sources. It can be demonstrated that with such a system in place and appropriate penalties for noncompliance, it is possible to achieve high rates of compliance even with tight monitoring and enforcement budgets.[49]

The necessary system is really quite simple, consisting only of an administrative tripartite grouping of pollution sources according to compliance status. The first group would consist of those sources that have been visited (or audited) and found to be in compliance. The probability of another visit to these sources might be made low enough that they would even feel they could relax their pollution control efforts. The second group would consist of those sources that have been found in violation at their last monitoring visit (if they were in the first group when visited) or that have just been "pardoned" or released from the third group. Sources in group two would not need to be audited any more frequently than sources in group one, but sources in group two that fail audits (that is, are found in violation) would be reclassified into group three, and would have to anticipate staying in that group for a long time. For the third group, audit visits would be conducted frequently—perhaps during every inspection period, but certainly often enough that it would not be worthwhile for any source in the group to attempt to get away with a violation. Thus being in group three would be equivalent to being in compliance. In such a system, properly designed, the vast majority of sources could be expected to end up in group two most of the time; and while in group two they would have an incentive to comply because of the threat of a long sentence to group three (rather than because of the immediate audit frequency for group three).[50]

The final recommendation is therefore:

- Under each major environmental regulatory program, establish a formal system linking expected frequency of monitoring visits (or audits of self-monitoring sources) to past records of discovered violation. The major requirement of such a system would be that sources found in violation twice in a row should be "sentenced" to frequent audits for a long period.

Together, adoption of these three suggestions would improve the efficiency of use of limited monitoring and enforcement budgets by increasing the probability that an actual violation would be discovered and by improving the way information about past compliance status is used to determine the level of current and future monitoring efforts. These recommendations are based on an assumption that despite the depressing evidence of widespread lack of compliance with strict environmental regulations, better compliance really is the goal of policy. The alternative possibility—that the legislation and accompanying regulations are meant to give the *appearance* of strictness while the reality is reflected by a lack of commitment to monitoring and enforcement—can only be ruled out by a future record that is an improvement on that of the past. The ongoing transition from absorption with initial compliance of conventional point sources of pollution to continuing compliance of hundreds of thousands of generators of a wide variety of wastes will require some fundamental redirection of thinking and effort on the part of the Environmental Protection Agency and state agencies. The three recommendations made here are meant to set the stage for that redirection and give it the greatest chance of success, given that substantial additional money is unlikely to be found in the midst of wars over budget deficits.

NOTES

1. Environmental Protection Agency, "A Framework for Statute-Specific Approaches to Penalty Assessments: Implementing EPA's Policy on Civil Penalties," EPA General Enforcement Policy no. GM-22, February 16, 1984.

2. See, for example, Robert W. Crandall and Paul R. Portney, "Environmental Policy," in Paul R. Portney, ed., *Natural Resources and the Environment: The Reagan Approach* (Washington, D.C., Urban Institute, 1984), especially pp. 49–52.

3. The General Accounting Office (GAO) has published two reports critical of ambient air quality monitoring; see GAO, "Air Quality: Do We Really Know What It Is?" GAO no. CED-79-84, May 31, 1979; and GAO, "Problems in Air Quality Monitoring System Affect Data Reliability," GAO no. CED-82-101, September 22, 1982.

4. The second judgment might turn out to be wide of the mark because in the actual regulatory system the connections made between source discharges and ambient quality results have generally been based on extremely simple models of complex natural systems and events.

5. Cheryl Wasserman, "Improving the Efficiency and Effectiveness of Compliance Monitoring and Enforcement of Environmental Policies: United States: A National Review,"

prepared for the Organization for Economic Co-operation and Development, Environmental Directorate, October 1984.

6. This point is made by Wasserman, albeit in a rather circumspect way. See Wasserman, "Improving the Efficiency and Effectiveness of Compliance Monitoring," pp. VI–5 through VI–7.

7. Clifford S. Russell, "Pollution Monitoring Survey: Summary Report" (Washington, D.C., Resources for the Future, 1983). This mail survey was sent to all 50 states, in some cases to more than one agency within a state. Responses were received from 27 states for air pollution questions and from 36 states for water pollution questions. For 21 states, responses covered both air and water pollution.

8. "New Groundwater Compliance Order Guidance Stresses Phased-in Remedies," *Inside EPA*, September 20, 1985, pp. 4,5.

9. The term audit has commonly been reserved by the EPA to refer to more general compliance checks. One of the enthusiasms of the early 1980s was that of self-audits, in which sources would check themselves over and assert their compliance in return for a reduction in actual agency monitoring.

10. Russell, "Pollution Monitoring Survey," p. 10.

11. This characteristic of standard pollution control monitoring stands in marked contrast to the system run by the Nuclear Regulatory Commission (NRC) to check on the performance of nuclear power plants. In the NRC system, each plant has a resident inspector (during both construction and operation) with freedom to go anywhere in the plant at any time, subject only to the same safety rules that apply to plant personnel. This inspector is encouraged to make rounds at random times and in varying order to add some element of surprise to the knowledge of routine monitoring. The resident inspector is supplemented by outside inspection teams that are encouraged to make their visits unannounced (Nuclear Regulatory Commission, Office of Inspection and Enforcement, *Inspection and Enforcement Manual*, chapter 0300 [January 1983].)

12. Whether or not the EPA and state pollution control agencies must have warrants for monitoring visits has not yet been considered by the U.S. Supreme Court, but the circuit and district courts that have reached the issue have made an analog between the EPA's scheme and that of the Occupational Safety and Health Administration and have thus required warrants. See *Public Service Company of Indiana, Inc. v. U.S. E.P.A.*, 509 F. Supp. 720 (S.D. Ind. 1981); *U.S. v. Stauffer Chemical Company*, 511 F. Supp. 744 (M.D. Tenn. 1981); *Bunker Hill Company v. U.S. E.P.A.*, 658 F. 2d 1280 (9th Cir. 1981). See also Robert W. Martin, Jr., "EPA and Administrative Inspections," *Florida State University Law Review* vol. 7 (Winter 1979) pp. 123–137. The courts have differed as to the type of warrants required for EPA inspections and the level of probable cause required to support them. Several have held that an ex parte warrant is all that is required; see *Bunker Hill Company v. U.S. E.P.A.*, 658 F. 2d at 1285; *U.S. v. Stauffer Chemical Company*, 511 F. Supp. 749. The Tenth Circuit reached the opposite conclusion in *Stauffer Chemical v. EPA*, 647 F. 2d 1075 (10th Cir. 1981). (An ex parte warrant may be granted at a hearing at which the party to be subject to search is not represented.)

13. "Supreme Court Roundup," *New York Times*, June 11, 1985; "Supreme Court Backs the EPA in Dow Dispute," *Wall Street Journal*, May 20, 1986.

14. See, for example, "EPA Will Tolerate Violations from 20% to 40% Over Water Permit Levels," *Inside EPA*, May 11, 1984, p. 7. This article reports on the contents of a draft definition of significant noncompliance then making the rounds within the agency.

15. General Accounting Office, *Waste Water Dischargers Are Not Complying with EPA Pollution Control Permits*, RCED-84-53 (Washington, D.C., 1983).

16. "Ruckelshaus Worried Citizen Suits Will Reveal Poor Enforcement Record," *Inside EPA,* May 11, 1984, p. 1.

17. "District Court Grants Judgment in Citizens Suit on NPDES Violation," *Inside EPA,* November 30, 1984, p. 5; "Court Diverges from Other Decisions in Siding with Citizens in CWA Suits," *Inside EPA,* August 30, 1985, p. 10.

18. Closer but far from identical. The probability of being found in violation, given that the decision to violate is made, is also a necessary deflater in the process of arriving at the expected penalty facing the source when it is deciding whether or not to violate intentionally. This probability cannot be estimated from available data.

19. Environmental Protection Agency, "Policy on Civil Penalties," EPA General Enforcement Policy no. GM-21, February 1984.

20. Genevra Richardson, "Policing the Enforcement Process," and Michael Hill, "The Role of the British Alkali and Clean Air Inspectorate in Air Pollution Control," both in *Policy Studies Journal* vol. 11, no. 1 (September 1982), pp. 153–164 and 165–174, respectively; Genevra Richardson, Anthony Ogus, and Paul Burrows, *Policing Pollution* (Oxford, Clarendon Press, 1982); Keith Hawkins, *Environment and Enforcement* (Oxford, Clarendon Press, 1984).

21. These studies have been summarized by the EPA in two volumes: James S. Vickery, Lori Cohen, and James Cummings, *Profile of Nine State and Local Air Pollution Agencies* (Washington, D.C., Environmental Protection Agency, 1981); Robert G. McInnes and Peter H. Anderson, *Characterization of Air Pollution Control Equipment Operation and Maintenance Problems* (Washington, D.C., Environmental Protection Agency, 1981).

22. Based on McInnes and Anderson, *Characterization of Air Pollution Control Equipment,* appendix A. The 10 percent figure has been calculated as the total excess emissions during violations, divided by the total allowable emissions, both in tons per year.

23. McInnes and Anderson, *Characterization of Air Pollution Control Equipment.* It should be noted that this summary contains some misleading numbers, apparently based on averaging procedures poorly chosen for the problem, though the text's explanation is far from clear on what was actually done with the raw data. The numbers reported above as relevant data were calculated directly from the data appendix provided.

24. General Accounting Office, *Improvements Needed in Controlling Major Air Pollution,* CED-78-165 (Washington, D.C., 1979).

25. General Accounting Office, *Air Pollution: Environmental Protection Agency's Inspections of Stationary Sources,* GAO/RCED-86-1BR (Washington, D.C., 1985).

26. General Accounting Office, *Costly Waste Water Treatment Plants Fail to Perform as Expected* (Washington, D.C., 1980); General Accounting Office, *Waste Water Dischargers Are Not Complying.*

27. The 1980 report is based on twelve months of data from 1978 and 1979, while the 1983 report uses data from October 1980 to March 1982, a period of eighteen months.

28. "Reporting" is a key word here. It is possible, for example, that because of different enforcement attitudes toward industrial and municipal sources, the former have incentives to shade reports toward compliance. This explanation of the contrast in the data competes with the explanation that compliance is actually better among industrial sources. In the absence of independent auditing, there is no way of choosing between the alternatives.

29. General Accounting Office, *Waste Water Dischargers Are Not Complying,* p. 25.

30. The requirement that continuing compliance monitoring (an inspection and maintenance program) be instituted applies to states that requested extensions beyond the 1982 deadlines for compliance with carbon monoxide and ozone NAAQSs. Extensions were

granted only when states could demonstrate that "all reasonably available control measures," if fully implemented, would not allow for attainment. In such situations implementation of, or at least a schedule for, an inspection and maintenance program was required before an extension could be approved when the area involved was urban and had a population greater than 200,000.

31. General Accounting Office, *Better Enforcement of Car Emission Standards—A Way to Improve Air Quality*, CED-78-180 (Washington, D.C., 1979).

32. General Accounting Office, *Vehicle Emissions Inspection and Maintenance Program Is Behind Schedule*, RCED-85-22 (Washington, D.C., 1985). Four states were on compliance schedules in this regard, but the latest agreed-to date for bringing a system in was early 1986.

33. General Accounting Office, *Vehicle Emissions: EPA Response to Questions on Its Inspections and Maintenance Program*, RCED-86-129 BR (Washington, D.C., 1986).

34. William L. Rosbe and Robert L. Gulley, "The Hazardous and Solid Waste Amendments of 1984: A Dramatic Overhaul of the Way America Manages Its Hazardous Wastes," *Environmental Law Reporter* vol. 14, no. 12 (1984) pp. 10458–10467.

35. General Accounting Office, *Illegal Disposal of Hazardous Wastes: Difficult to Detect or Deter*, RCED-85-2 (Washington, D.C., 1985). The manifest requirement became effective in November 1980.

36. Community service requirements were the only penalty imposed in 4 cases; and one conviction resulted only in an order to comply in the future. Ibid., pp. 39, 40.

37. General Accounting Office, *Hazardous Waste: Facility Inspections Are Not Thorough and Complete*, RCED-88-20 (Washington, D.C., 1987).

38. Another GAO report found that the EPA was doing a reasonably good job of checking up on land disposal facilities that had certified their compliance with requirements for groundwater monitoring and financial responsibility (such certification was required to allow continued operation of the facilities after November 8, 1985). Similarly, the EPA was found to have made an effort to check up on the facilities that did not certify compliance and thus were required to cease operating (General Accounting Office, *Hazardous Waste: Enforcement of Certification Requirement for Land Disposal Activities*, RCED-87-60 BR [Washington, D.C., 1987]).

39. "Alm Forms Task Force to Recommend Ways to Improve Monitoring," *Inside EPA*, September 30, 1983, p. 14.

40. Rochelle Stanfield, "Ruckelshaus Casts EPA as 'Gorilla' in States' Enforcement Closet," *National Journal*, May 26, 1984, pp. 1034–1038.

41. Ibid.

42. Environmental Protection Agency, "Policy on Civil Penalties"; and Environmental Protection Agency, "A Framework for Statute-Specific Approaches to Penalty Assessments."

43. Tamar Levin, "Business and the Law," *New York Times*, March 5, 1985. Levin quotes James Oppliger, a deputy district attorney general in California, as saying, "A person facing jail is the best deterrent against wrongdoing."

44. Daniel Riesel, "Criminal Prosecution and Defense of Environmental Wrongs," *Environmental Law Reporter* vol. 15, no. 3 (1985) pp. 10065–10081. Riesel provides some historical background to this practice, but is mainly concerned with a scholarly review of the applicable law.

45. Barry Meier, "Against Heavy Odds, EPA Tries to Convict Polluters and Dumpers," *Wall Street Journal*, January 7, 1985.

46. "Justice Environmental Crimes Unit Wins Record Number of Indictments," *Inside EPA*, September 6, 1985, pp. 5,6.

47. See, for example, Roy J. Harris, Jr., "Mobile Plant Raided in Investigation of Pollution Charge," *Wall Street Journal,* July 29, 1985, describing a criminal investigation raid by Los Angeles County officials; Victoria Churchville, "Executive on Trial in Harbor Dumping Case," *Washington Post,* September 5, 1985, which relishes the contrast between the defendant executive's appearance and his role as an accused criminal; and Tom Vesey, "D.C. Man Jailed for Polluting," *Washington Post,* September 10, 1985, which notes that the defendant, owner of a small dump site, actually is serving time in jail.

48. See Judson W. Starr, "Environmental Enforcement Takes an Ominous New Turn: Managers and Officers Go Directly to Jail," *Environmental News* vol. 2, no. 1 (Fall 1988) pp. 1–3 (publication distributed by Venable, Baetjer, and Howard, Baltimore).

49. See Clifford S. Russell, "Game Theory Models for Structuring Monitoring and Enforcement Systems," *Natural Resource Modeling,* forthcoming. The same end could be achieved by using differential penalties, dependent on past compliance record, for what is necessary is that repeat offenders face very high *expected* penalties. But there is some practical reason to prefer a system that operates through the probability of a monitoring visit, because very high penalties may never be assessed on individual violators.

50. In a simple world without measurement error it would be possible to describe the system more neatly, and the sentence to group three would become indefinite or infinite and no rational source would ever end up in that group. Since measurement error is inevitable, even if all sources try to avoid group three, some will fail and will be sentenced to a term there. If all such terms were indefinite, all sources would eventually be included in group three and the system would break the budget constraint by requiring too many audits per period. Thus it is necessary that some fraction of the sources in group three be pardoned each period. The fraction must be small enough that the sources expect to spend a long time in compliance. How long is long depends on other characteristics of the problem, including the size of fines for violations relative to the costs of compliance.

eight

Overall Assessment and Future Directions

Paul R. Portney

Two questions might occur to the reader who has come this far: What overall conclusions may we draw about the success of federal environmental regulation in the United States through the beginning of the 1990s? What, if anything, can we say about the possible future direction of environmental policy in this country in the years to come? The remaining pages of this book are devoted to these two questions.

PROGRESS TO DATE

As the preceding pages have shown, there are several ways by which one might measure progress in environmental policy. These include number of regulations issued, control measures put in place, enforcement actions taken, and, finally, actual physical improvements in environmental quality. It should be clear from the earlier chapters that, where it is applicable,[1] this last measure is by far the most important. After all, it does little good to issue regulations on time, ensure that affected sources comply promptly with them, and penalize those which do not if this flurry of activities does little to improve environmental conditions along one dimension or another.

Environmental Improvements

When evaluated in this way—according to demonstrable environmental results—the record of federal environmental policy is a mixed one. Only in the case of air quality can widespread improvements be shown. There we have real reason to be encouraged. For most major metropolitan areas, ambient concentrations of almost all the common air pollutants have declined; this is particularly true for lead, sulfur dioxide, and particulate matter, three of the most worrisome pollutants from the standpoint of human health. Even in those areas where improvements have been slow or even nonexistent, one can fairly argue that things would be worse still were it not for nearly two decades of investments in pollution control required of automakers, industrial sources, municipalities, and others. While we have no widespread monitoring network to provide confirmatory evidence, it is likely that this prolonged period of controlling the common (or criteria) air pollutants has also resulted in substantial reductions in airborne concentrations of many less common hazardous air pollutants (see chapter 3 concerning the distinctions).

As always, we must hedge our conclusions somewhat. First, ozone continues to be a stubborn air quality problem in many areas and will continue to be so for some time to come. It has proved far more difficult to control than the pollutants mentioned above. Second, we must remember that at least some of the improvements in air quality cited here and in chapter 3 are due to shifting patterns of industrial activity. To put it starkly, the slow and painful declines of the domestic steel, mining, and other heavy industries have contributed to the improvement in air quality in some areas. We would prefer that our environmental improvements not come at the expense of such losses.

It is hard to be as optimistic about the progress made in other environmental areas. As chapter 4 points out, improvements in water quality have been much more sporadic and uneven. In some locations, particularly in urban areas where water quality had deteriorated badly throughout the 1950s and 1960s, dramatic turnarounds have taken place. Fishing and swimming are now possible in some places where they were unthinkable even ten years ago.

Yet such success stories are idiosyncratic rather than widespread. Moreover, water quality has deteriorated in many areas since 1972, when the federal government greatly stepped up its presence in water pollution control. In part this has been due to a reluctance on the part of Congress and the Environmental Protection Agency to take on the problems associated with non-point sources of water pollution, such as farms and the streets and parking lots in urban areas. Quite frankly, it is harder to whip up opposition to such sources than to industrial sources, and as a result their share of the overall water pollution picture has increased over time. Additional progress in water pollution control will depend in large part on summoning the political will and financial resources necessary to address these polluters.

It is premature to attempt any definitive evaluation of progress in cleaning up abandoned and active hazardous waste disposal sites. After all, the basic statutes—RCRA, enacted in 1976, and CERCLA (or Superfund), enacted in 1980—have not been in place as long as our air and water pollution control laws, and they were slower to be implemented once they were passed. Nevertheless, some tentative inferences are appropriate, and they suggest that all is not well.

To this point, at least, it would appear that the major impact of RCRA regulations (which deal in large part with currently operating hazardous waste disposal sites) has been to greatly reduce the number of such facilities. As chapter 5 points out, the number fell by nearly two-thirds following regulations stipulating that such facilities must have liability insurance to cover any clean-up costs that might be required following their eventual closure. While this makes it much more difficult to find permitted disposal sites, clearly such pruning is not all bad. To the extent the insurance requirement has weeded out sites which were operating in a careless fashion, it is encouraging to know that wastes will no longer be deposited there. On the other hand, we now must face the task of determining what remedial measures must be taken at those sites which did close, lest they find their way onto the National Priorities List (as some already have done).

It is much more difficult to ascertain the effects of RCRA regulations on those disposal sites that have continued to operate. To be sure, these facilities face new requirements concerning the technological safeguards that must be in place at each site, the groundwater monitoring they must do, and the reporting they must do in concert with those who generate, store, and/or transport hazardous wastes to the sites. In the long run, perhaps the most important feature of RCRA will be the phased prohibitions on land disposal of certain classes of hazardous wastes. Once the EPA determines which wastes can and cannot be safely disposed of even in secure landfills, other disposal options must be identified, or the generation of such wastes must be reduced. It will not be easy to find alternative means of disposal, since incineration—the most obvious alternative—is likely to be resisted by parties concerned about its risks or visual disamenities (see chapter 5).

Superfund may one day lend itself to easier evaluation. The program was put in place to clean up abandoned hazardous waste disposal sites, so we should be able to mark progress in part by looking at the number of cleanups that have taken place. A review of the record to date is not encouraging. Although the Superfund (CERCLA) Act was passed in 1980, the EPA has reported that through March 1989 work had been completed at only 41 sites, only 26 of which had been removed from the National Priorties List. Since the NPL now lists nearly 1,200 sites, and since the list of potential candidates for the NPL ranges from 30,000 to 400,000 (see chapter 5), the task ahead is sobering.

On the other hand, we must not fall into the trap of counting only finished cleanups. Emergency removals—the EPA's initial response upon ascertaining that a threat to health may exist at a site—have been much more frequent, numbering more than 1,300, counting actions at NPL and other sites. As chapter 5 suggests, some of these actions may remove the lion's share of the risks associated with a particular site; in fact, at some sites it might be wise to take no further action after an emergency removal has taken place, although the law makes it difficult to stop at that point.

The most difficult evaluative assignment concerns our laws governing pesticides and toxic substances and the regulations written pursuant to them. First of all, these laws are not directed at pollutants in the air or water or at the waste dumps that dot our landscape. Rather, they are intended to increase the information we possess about the pesticides and toxic substances society uses in a variety of ways. They are aimed at products rather than waste streams, in other words.

Nevertheless, as chapter 6 points out, progress under both the Toxic Substances Control Act and the Federal Insecticide, Fungicide, and Rodenticide Act has been less than overwhelming. While new chemicals and pesticide introductions do include more testing data and other kinds of information than before, our record for testing products already in commerce at the time the laws were passed is not good. This is due to reasons both internal and external to the Environmental Protection Agency. As to the former, it has simply taken the EPA too long to gear up the TSCA program. On the other hand, certain features of TSCA—its requirement that any testing rules be promulgated as formal regulations, for instance—are not conducive to streamlined information-gathering. It will always be more difficult to keep score on TSCA and FIFRA than on the other environmental statutes.

Turning to environmental enforcement efforts, we encounter difficulties similar to those plaguing efforts to evaluate our pesticide and toxic substance laws. As chapter 7 makes clear, we have graduated in enforcement from worrying about forcing the installation of control equipment at regulated sources; we are now concerned with the more important but also more difficult problem of ensuring that that equipment is operated correctly and routinely. Unfortunately, for evidence on these important issues we often rely on information provided by the regulatees themselves (the environmental equivalent of the familiar problem of the fox guarding the chicken coop).

While scanty, the empirical evidence that does exist tells a mixed story. Some audits suggest that compliance with some standards is good; for example, a high percentage of industrial water polluters were found to be in compliance with the first-stage requirements issued under the Clean Water Act. On the other hand, municipal waste treatment plants—major water polluters in their own right—were found in one major study not only to be out of compliance often, but also to be significantly so for alarmingly long periods of time. Other enforcement areas of concern include (1) the performance of motor vehicles (the pollution control devices on which often deteriorate rapidly); (2) the persistence of violations at some sources which have repeatedly been cited for noncompliance, suggesting a lack of political will; and (3) poor compliance with groundwater monitoring and other requirements at hazardous waste disposal sites.

An Environmental Consciousness

Although it cannot be measured like physical changes in air or water quality, or the number of hazardous waste sites, in one other respect federal environmental policy has had an important impact on life in the United States: I refer to the development of an environmental consciousness (or ethic) which colors virtually all decisions made by corporations, by federal, state, and local government operating units, and by many ordinary citizens. Environmental consciousness is not unrelated to quantitative measures of progress; indeed, without it less progress would have been made over the last twenty years than is reported above. Nevertheless, it is worth mentioning in its own right.

This consciousness is perhaps most clearly reflected in public opinion polls. While varying somewhat over time on the specifics, the polls consistently indicate that the environment, broadly defined, is near the top of the list of public concerns. Even when environmental improvements are described as involving economic sacrifices, there is strong support for pursuing them.[2] As a result, it is now second nature for the public to inquire about the environmental consequences of large public or private development projects and, moreover, to object loudly and strenuously if it is dissatisfied with the answers it gets. It is interesting, however, that environmental consciousness is often less heightened when it comes to decisions made at the individual level. As chapter 5 points out, few people realize that the paint thinners and used motor oil they often pour down their drains, or the batteries they discard in their garbage, are every bit as hazardous as some of the wastes generated in the manufacturing sector. Real progress in reducing the volume of solid and hazardous wastes will require effective individual as well as collective action.

The environmental ethic described here has had an unmistakable effect on corporate decision makers. In part this has been a matter of necessity. Virtually every corporate decision—whether concerning the construction of a new plant, the modernization or closure of an existing one, the introduction of a new product line, or the modification of a current product—requires a careful analysis of the regulatory implications. But the heightened sensitivity regarding the environment is not altogether forced, as a recent example indicates.

In the 1986 amendments to the Superfund law, Congress tacked on a separate statute, the Emergency Planning and Community Right-to-Know Act of 1986. Among other things, this law required that for the first time certain classes of manufacturing facilities report annually their emissions of a large number of toxic chemicals into the air, water, and land, even if these emissions were routine and allowable under all other environmental statutes. The statute was designed primarily to make information available to state and local officials, community groups, and individual citizens. When in mid-1988 the chairman of the board of the Monsanto Company saw the volume of emissions his company reported, he immediately pledged to reduce those emissions by 90 percent over the next five years. He did this in spite of the fact that Monsanto was violating no laws with those emissions and that such an ambitious cutback would be quite expensive, expressing the view that Monsanto could and should do better—and would. While one should not read too much into such decisions, they are not altogether uncommon, and they reflect an enhanced sense of environmental responsibility on the part of many large corporations. This, too, is a very real accomplishment of the last two decades of environmental legislation and regulation.

Economics in Environmental Policy

The authors of the preceding chapters have not hidden their belief that economics, if carefully used, can play a valuable role in improving environmental regulation. It is fair to ask, then, for a summary judgment on its actual role.

As is pointed out in the introductory chapter, the evidence is mixed in regard to the statutory role allowed for economic considerations in environmental regulation. Two important laws—the Toxic Substances Control Act and the Federal Insecticide, Fungicide, and Rodenticide

Act—*require* the administrator of the EPA to balance the risks against the economic benefits that attend the chemicals used in commerce or the pesticides and herbicides used in agriculture. One reason for this explicit balancing approach is clear: the risks that arise from the chemicals or pesticides in question are believed to be inherent in the products themselves.

In contrast, other environmental risks are viewed as arising from the *production* (rather than the use) of steel, electricity, paper, or other products. In other words, many of the risks are associated with residual by-products of production, and the approach has been to write laws and regulations aimed at reducing (or even eliminating) these by-products (see chapter 7). Thus the laws governing air and water pollution and hazardous waste disposal are aimed at the pollution itself; these laws leave relatively little room for a balancing of benefits and costs. Rather, they tend to embrace a technology-based approach, often coupled with mandates for ambient standards that provide margins of safety.

While perhaps understandable, the distinction between risks that are inherent in product use and those that result from the production of certain products is quite artificial. One should probably view a ton of steel (or other manufactured products) as possessing certain desirable characteristics, as well as a set of undesirable attributes—namely, the pollution that results when it is produced. While the undesirable characteristics can be mitigated, perhaps significantly so, they cannot be completely eliminated. In this respect, then, these manufactured products pose problems analogous to the products regulated under TSCA or FIFRA. An appropriate response might be to allow the same kind of balancing in standard-setting that those latter statutes allow. However logical this may seem, it has not been the approach taken in air and water pollution control or hazardous waste management. In fact, in mid-1989 legislation was introduced in Congress which, if passed, would remove the balancing mandate in FIFRA and make it a risk-based statute alone.

It would be misleading to conclude on this note, however. While the statutes that govern air and water pollution control, as well as hazardous waste regulation, do not mandate economic or balance-oriented approaches, they generally do not forbid them in bold letters. Thus room is left for interpretations of the statutes that are more receptive to economic considerations. Consider, for example, the EPA's attempt to formulate a policy to regulate hazardous air pollutants under the Clean Air Act (see chapter 3). For a time the agency employed an approach in which any residual risks remaining after the application of technology-based standards were to be weighed by the individual states to determine whether they were worth controlling further. While this approach was eventually overturned by the courts, it did represent an effort to inject some trading-off of risk reductions against economic considerations. Such efforts will no doubt continue in the future, but they are a poor substitute for laws that explicitly require balancing-type approaches to regulation.

Will environmental law evolve to include a more prominent role for economic considerations? Writing from the vantage point of 1989, the answer would appear to be no. If anything, the more recent environmental laws—the 1980 Superfund act, the 1984 amendments to RCRA, and the 1986 SARA amendments to Superfund—move somewhat away from allowing economic considerations in standard-setting (see chapter 5). To take but one example, under SARA remedial actions at Superfund sites are to provide for "permanent remedies" for disposal of the hazardous materials regardless of the costs involved (although use of the least expensive technique to provide this absolute protection is encouraged).

There are several reasons for this. First—and quite frankly—economists have not done a convincing job of explaining why a balancing-type approach to environmental policy is advantageous. They are perceived as being blind (or at least myopic) to the real-world problems that would arise in implementing solutions that look attractive in theory.[3] In addition, most lawmakers prefer not to acknowledge openly that environmental protection must be traded off against economic considerations in the same way that human health, housing, national defense, education, and other important national objectives are balanced in federal budget policy each year. Since they know that these tradeoffs will be made implicitly at the Environmental Protection Agency even if the laws state otherwise, they see little to be lost by writing zero-risk laws.

FUTURE DIRECTIONS

Environmental policy in the United States is sure to change in unforeseen ways in years to come. At the same time, it is possible to identify three recent and important trends that are likely to influence the direction of change. These trends concern environmental federalism, the withering of administrative discretion, and the global dimensions of environmental pollution.

Environmental Federalism

Although the statutes discussed above (especially the Clean Air and Clean Water acts) often reserve important responsibilities for state and sometimes local governments, they are all federal laws. Yet there is no reason why environmental policy must be made at the federal level.[4] Over the past decade or so some of the most interesting environmental initiatives have arisen at the state level. This began with the individual states gradually taking over responsibility for the operation and management of air and water pollution programs under the federal laws. As chapter 4 indicates, nearly forty states now handle virtually all aspects of regulation under the Clean Water Act, and even more states have been delegated the responsibility of issuing permits under the Clean Air Act.

More important, the individual states have become active in passing their own environmental laws. For example, by early 1989 five states (New York, Wisconsin, Massachusetts, Minnesota, and New Hampshire) had passed laws to combat acid rain, and one state (Michigan) had tightened a previous emissions control law because of concern about acid deposition. Similarly, forty states now have their own versions of the federal Superfund law to provide monies to clean up abandoned hazardous waste disposal sites. Most significant, perhaps, has been the passage in California of the Safe Drinking Water and Toxic Enforcement Act of 1986. Better known as Proposition 65—its name on the November 1986 referendum ballot on which it appeared—the act sets in motion a process to prohibit certain discharges of toxic chemicals, and also requires that "clear and reasonable warnings" be given to anyone exposed to such toxic chemicals. Details surrounding the implementation of the act are only now being worked out. But it appears that this law may have as far-reaching an effect on emissions of air and water pollutants in California as the Clean Air or Clean Water acts. Other states have begun to evince interest in California's approach.

How should one view this resurgence of state regulatory activity? There are good arguments both for and against it. In the case of acid rain, for instance, actions by a handful of individual states are no substitute for a coherent national policy. Interjurisdictional externalities are simply too pervasive in this case for states to effectively act alone. In fact, acid deposition should probably be addressed at the international rather than the national level (see below). This is also true for many water pollution problems.

Also militating against decentralized environmental regulation is the possibility that it would foster an unhealthy competition between states to reduce environmental standards in order to attract new business growth. Although there is little empirical evidence to suggest that this has been so to date, it remains a serious concern to many. Beyond that, environmental standards that differ greatly from state to state would make it very difficult for companies doing business in many different areas. This is most obvious in the case of motor vehicle emissions standards; automakers simply could not manufacture cars to meet fifty different sets of standards.[5] As in this case, some sort of federal preemption may occasionally be required.

On the other hand, there are reasons to be encouraged about the proliferation of state initiatives. To begin with, while some environmental problems clearly transcend state (or even national) boundaries, others do not. Where they do not, decentralized regulation may reflect local conditions and tastes better than top-down regulation from Washington. There is no reason, for example, why the extent of clean-up activities at each abandoned hazardous waste disposal site should be the same from state to state. If no spillovers from one area to another exist (and they are unlikely to in the case of such sites), it might make sense to allow states or even local governments the right to decide how far each cleanup should proceed. Similarly, if some communities are willing to tolerate somewhat higher levels of certain contaminants in their drinking water than the federal government feels are advisable, it is not clear why they should be prohibited from doing so. Even if such decentralization in standard-setting were permitted, the federal government would still play the important role of sponsoring research on the adverse health and ecological effects associated with environmental pollutants, the control technologies that could be used to address certain types of problems, and so on.

Another reason for allowing some decentralization is more pragmatic. Since federal statutes already delegate most enforcement (as well as other) powers to the states, it can be argued that a sort of de facto decentralization already exists. The states that want a pristine environment will enforce the federal laws vigorously, while those more concerned about business growth may be more lax. If this is so, it *may* make sense to give the states more latitude to set the original standards. Regardless of how one views the merits of these competing arguments, however, it is clear that the individual states will continue to play an increasing role in the overall scheme of environmental policy.

The Withering of Administrative Discretion

Through the mid-1980s, Congress adhered to a fairly consistent approach in the environmental legislation it passed. That was to use the statutes to sketch out the broad directions Congress wanted federal environmental policy to take, but to delegate to the administrator and other officials of the EPA the responsibility to fashion the specific policies for carrying out the broader mandates. In fact, this was the approach underlying the creation of virtually all federal regulatory agencies (see chapter 2). In environmental legislation there were always conspicuous exceptions to the general rule, of course. For instance, Congress mandated specific motor vehicle emissions reductions in the Clean Air Act rather than delegating to the EPA the power to establish them. Similarly, in passing the Toxic Substances Control Act, Congress held that polychlorinated biphenyls (PCBs) be strictly regulated, but delegated to the EPA the responsibility to determine which other chemicals should be tested and controlled. Nevertheless, the general rule was to provide a good deal of discretion to the officials of the Environmental Protection Agency in choosing what to regulate and how and when to do it.

This is no longer the case. Beginning with the 1984 amendments to the Resource Conservation and Recovery Act, Congress began to reduce the flexibility previously granted regulatory officials. As chapter 5 points out, those amendments established so-called hammer provisions—congressionally mandated standards and/or bans that would automatically take effect unless the EPA could show why this should not be so. This erosion of administrative discretion continued with the 1986 amendments to the Safe Drinking Water Act (which decreed that the EPA regulate more than eighty specific contaminants in a three-year period), as well as in the 1986 Superfund Amendments and Reauthorization Act. And if the proposed amendments to the Clean Air Act discussed in late 1989 are any indication, the trend shows no signs of abating. Among other things, those proposed amendments would have required the EPA administrator to regulate a long list of hazardous air pollutants and to do so in a very specific way; currently, the EPA has the discretion to choose both the pollutants to target for regulatory activity as well as the approach to be taken.

There are several reasons for this new congressional posture. Most important, it was a reaction to the debacle at the EPA between 1981 and 1983, when the agency was staffed with officials in whom Congress (and, ultimately, the president who appointed them) had no confidence.[6] Fearing a lack of commitment by those officials to environmental protection, Congress became loath to give as much latitude to the EPA as it had in the past. Unfortunately, even after credibility was restored to the agency, Congress continued to reduce the discretion the agency had enjoyed in the past.

There are other reasons for the loss of discretion granted the EPA. One is the agency's record of missing deadlines established in the environmental statutes.[7] This has led some in Congress to believe that "default" provisions—standards that would automatically take effect unless the EPA issues its own regulations by certain deadlines—are essential to environmental protection. Also, the complexity of the issues with which the EPA deals has made effective congressional oversight much more difficult.[8] As a result, sentiment has grown for specifying in the laws themselves which substances are to be regulated, what means are to be used in doing so, and when this should be accomplished. Implicit in this approach is the belief that it will vitiate the need for time-consuming and rancorous oversight hearings in which Congress struggles to determine just what the EPA is doing and why.

While in part understandable, there are reasons to be concerned about this loss of discretion. First, it results in part from misplaced blame. The EPA has missed many deadlines, but as the previous chapters point out, these deadlines have often required the EPA to issue a great number of very complicated standards in a very short period of time. This simply is not reasonable. Moreover, Congress has consistently given the EPA more assignments than resources. As seen in chapter 2, the Environmental Protection Agency now has an operating budget that is considerably smaller in real terms than it was nine years ago, even though the agency has been given important new responsibilities during that time. This is a sure recipe for falling behind.

The most important reason to be concerned about this recent trend is that flexibility and discretion are key to really effective environmental management.[9] The problems that must be addressed by the modern EPA are simply too diverse and complex for the use of uniform, across-the-board solutions. Yet Congress has neither the time nor the inclination to provide other solutions. This means excessive control at many sources of pollutants even though special conditions, economic or otherwise, might militate against it; it also means that, to the extent affordability is a criterion for regulation, some sources that should be closed down because of the damage they do will remain open. While more flexibility surely does not guarantee enlightened regulation, the job is made more difficult when statutory straightjackets sharply delimit administrative discretion. One can only hope that the EPA—and for that matter other social regulatory agencies—regains the confidence of our national legislators. Until that time, it appears likely that Congress will play a larger and larger role in the design of specific environmental regulations.

Global Concerns

The late 1980s have brought with them a growing awareness of the global dimensions of some environmental problems. To be sure, the trans-boundary nature of many such problems had been recognized in the past. In fact, international agreements have been negotiated on traditional air and water pollution problems, trade in endangered species, commercial fishing and whaling, and other environmental matters. Nevertheless, domestic problems have always attracted the most attention. This may no longer be the case. Recently, public attention has been focused on two less traditional but potentially more serious environmental difficulties: the breakdown of the protective layer of stratospheric ozone, and the accumulation in the atmosphere of carbon dioxide and other so-called greenhouse gases that could significantly increase global temperatures and change climate patterns.

Growing awareness of these problems both led to and was heightened by the signing in 1987 of an international accord to limit worldwide emissions of chlorofluorocarbons (CFCs). These man-made chemicals have the potential to break down the protective layer of stratospheric ozone that helps shield us from damaging ultraviolet radiation from the sun. They also act as greenhouse gases. Although the United States and a few other countries had previously banned all aerosol uses of CFCs because of concern over ozone depletion, the so-called Montreal Protocol will necessitate significant further restrictions on CFC production and use among all signatories.

Public concern has also become heightened about global warming. Carbon dioxide comes mainly from the combustion of coal, oil, wood, and other fuels, while the other trace gases come from a diverse set of sources. These gases allow solar radiation to reach the earth, but can trap the outgoing terrestrial radiation and thus contribute to global warming.[10] It cannot yet be demonstrated that this process has resulted in warmer average global temperatures. Nevertheless, over the last 100 years, at least, temperatures appear to have increased; so, too, have atmospheric concentrations of carbon dioxide and, probably, other trace gases. Thus the evidence is consistent with the warming hypothesis. Because warmer temperatures and the altered weather patterns that would accompany them could have a number of quite disruptive effects (altered energy requirements, shifts in crop patterns, and particularly increases in sea level), concern has grown.[11]

In a sense, ozone depletion and global warming are not only difficult to understand, but also difficult to fit into a discussion of traditional environmental policy.[12] For one thing, the pollutants of concern—CFCs and carbon dioxide—do little or no direct harm to human health and the environment. In this regard, they are unlike the air and water pollutants, hazardous wastes, or industrial chemicals discussed in the previous chapters. For this reason, it is arguably more difficult to address them under the present statutory framework.[13] Also, the uncertainties involved in assessing the seriousness of these problems, not to mention fashioning appropriate policy responses, is much greater than for many conventional environmental problems. And finally, in the case of global warming, possible responses transcend traditional regulatory measures. For instance, reducing emissions of greenhouse gases would require major changes in the levels and patterns of energy use in the United States and elsewhere, as well as reductions in the rate at which tropical rainforests are harvested around the world. Neither lend themselves easily to the type of control measures used by the EPA.

It is hard to say how long public attention will be directed at these two problems. As it accumulates, scientific evidence may reinforce current concerns; if so, interest is likely to remain high or grow. On the other hand, new knowledge may reassure rather than alarm us further. In this case, other and perhaps more traditional problems may come to occupy center stage. In one regard this would be unfortunate. Although some environmental problems do have fairly narrow boundaries, this is less so than ever before. As the world economy becomes increasingly integrated, the effects of domestic fiscal and monetary policies will be felt elsewhere; so, too, will serious environmental problems in the United States, Japan, Europe, and South America come to be felt all around the world. One can only hope that this will spur more international cooperation in preventing such problems before they occur.

NOTES

1. As chapter 6 points out, under the pesticide and toxic substance laws the purpose of the regulations is to provide information about the toxicities of chemical constituents and products so that harmful substances might be prevented from coming to market or, if already there, might be withdrawn. Thus these regulatory programs have a different focus.

2. For a survey of these findings, see Dunlap (1987).

3. Kelman (1981) presents an interesting discussion of the difficulties economists have had in making a case for effluent charges or other incentive-like approaches to environmental regulation. For a recent and quite sober assessment of the potential for these approaches to regulation, see Tietenberg (1985).

4. For an excellent discussion of the appropriate roles of the various levels of government in environmental policy, see Congressional Budget Office (1988).

5. As pointed out in chapter 3, under the Clean Air Act California is allowed to maintain vehicle emission standards tighter than those for the rest of the country, because that state already had such standards on the books when the act was passed.

6. See Crandall and Portney (1984) for a discussion of these problems. Florio (1986) discusses the special problems that arose under RCRA and Superfund.

7. This record is discussed at length in Environmental and Energy Study Institute (1985); see also Florio (1986).

8. See National Academy of Public Administration (1988) for a discussion of these difficulties.

9. Even proponents of the new congressional activism acknowledge this serious shortcoming in the congressional approach. See Florio (1986), especially p. 379.

10. See Rosenberg (1988).

11. Some of the effects could be benign, however. This may be obvious to anyone who must regularly endure midwestern winters!

12. These two problems are described at some length, and policy responses are considered, in Firor and Portney (1982).

13. The Clean Air Act does contain a provision enabling the EPA to regulate CFCs.

REFERENCES, chapter 8

Congressional Budget Office. 1988. "Environmental Federalism: Allocating Responsibilities for Environmental Protection," staff working paper, Natural Resources and Commerce Division.

Crandall, Robert, and Paul R. Portney. 1984. "Environmental Policy," in Paul R. Portney, ed., *Natural Resources and the Environment: The Reagan Approach* (Washington, D.C., Urban Institute) pp. 47–81.

Dunlap, Riley E. 1987. "Public Opinion on the Environment in the Reagan Era." *Environment* vol. 9, no. 6, pp. 7–37.

Environmental and Energy Study Institute. 1985. *Statutory Deadlines in Environmental Legislation: Necessary but Need Improvement* (Washington, D.C.).

Firor, John W., and Paul R. Portney. 1982. "The Global Climate," in Paul R. Portney, ed., *Current Issues in Natural Resource Policy* (Washington, D.C., Resources for the Future).

Florio, James J. 1986. "Congress as Reluctant Regulator: Hazardous Waste Policy in the 1980s," *Yale Journal on Regulation* vol. 3, no. 2, pp. 351–382.

Kelman, Steven. 1981. *What Price Incentives?* (Boston, Auburn House).

National Academy of Public Administration. 1988. *Congressional Oversight of Regulatory Agencies: The Need to Strike a Balance and Focus on Performance* (Washington, D.C., NAPA).

Rosenberg, Norman J. 1988. *Climate Change: A Primer* (Washington, D.C., Resources for the Future).

Tietenberg, T. H. 1985. *Emissions Trading: An Exercise in Reforming Pollution Policy* (Washington, D.C., Resources for the Future).

Index